O9-BUA-277

THE ILLUSTRATED ENCYCLOPEDIA OF
MAMMALS

THE ILLUSTRATED ENCYCLOPEDIA OF MAMMALS

By Dr Vladimír Hanák
and Dr Vratislav Mazák

Edited by
Susan Lowry

CHARTWELL
BOOKS, INC.

PICTURE ACKNOWLEDGEMENTS

The figures indicate the numbers of photographs

Anděra, M.: 2, 38, 39, 44, 46, 54, 60, 61, 139, 140, 145, 146, 152, 153 **Archives:** 15 **Assen van, R.:** 412 **Bailey, J. A.** (Ardea): 217 **Bárta, D.:** 84, 85, 161, 182, 183, 189, 191, 201, 202, 208, 221, 240, 261, 285, 287, 315, 317, 324, 325, 332, 336, 360, 363, 396, 400 **Bartlett, J. & C.** (Coleman): 141 **Bartoš, L.:** 43, 79, 106, 126, 193, 222, 283, 335, 338, 361, 392, 395, frontispiece **Beames, I. R.** (Ardea): 129 **Bisserot, S. C.** (Coleman): 50, 57, 77 **Blahout, M.:** 386, 389 **Borland, R.** (Coleman): 41 **Boulton, M.** (Coleman): 96 **Burton, J.** (Coleman): 8, 51, 56, 159 **Carr, R.** (Coleman): 136 **Castel, P.:** (Jacana): 265 **Coleman, B.** (Coleman): 166 **Col. Varin – Visage** (Jacana): 419 **Devez CNRS** (Jacana): 37 **Dobroruka, L. J.:** 203 **Erize, F.** (Coleman): 26, 262, 414 **Figala, J.:** 121, 133, 138 **Foot, J.** (Coleman): 264 **Giddings, A.** (Coleman): 263 **Groves, C. P.:** 229, 231, 232, 238, 246 **Haleš, J.:** 42 **Heráň, I.:** 257 **Heurn, van:** 422 **Hodková, Z.:** 128 **Janou** (Jacana): 134 **Judin:** 40, 49, 124, 127, 303 **Klemm, de Ph. C.** (Jacana): 322 **Kunc, Z.:** 241 **Laycock, G.** (Coleman): 267 **Mazák, V.:** 176, 223, 242, 255, 274, 286, 288, 333, 345, 399, 408, 423 **Mohrová, E.:** 268, 269 **Müller, E.:** 27, 218 **Myers, N.** (Coleman): 69 **Okapia:** 28, 67, 397 **Opatrný, E.:** 20, 175, 186, 190, 197, 198, 199, 200, 237, 319, 320, 379, 391, 394, 398, 406, 410 **Paul, P.:** 62 **Power, A.** (Coleman): 292 **Pucholt, R.:** 73, 99, 111, 177, 185, 188, 204, 205, 207, 211, 266, 298, 321, 323, 326, 393 **Riewe, R.:** 306, 413, 418 **Rödl, P.:** 45, 59, 118, 144, 147, 150, 154, 164, 168, 170, 172, 180, 195, 304, 307 **Roer:** 63 **Rys, J.:** 53, 58, 122, 131, 137, 142, 148, 149, 155, 157, 165, 169, 171, 174, 308, 337, 388 **Scholz, K.** (ZEFA): 119 **Seget, J.:** 9, 248, 250, 251, 258, 280, 401 **Smola, O.:** 196, 215, 224, 295, 349, 355, 383, 384, 385 **Staněk, V. J.:** 4, 52, 55, 70, 72, 93, 100, 101, 102, 103 104, 110, 113, 116, 123, 156, 160, 162, 163, 230, 236, 243, 245, 305 **Sundance, J. X.** (Jacana): 184 **Thau, A.** (ZEFA): 10 **Trevor, S.** (Coleman): 296 **Vágner, J.:** 87, 130, 212, 228, 249, 252, 253, 254, 256, 272, 273, 275, 276, 277, 289, 290, 297, 311, 312, 316, 331, 334, 341, 342, 344, 346, 347, 348, 350, 351, 352, 353, 354, 357, 358, 362, 364, 365, 366, 367, 368, 369, 371, 372, 373, 374, 375, 376, 377, 378, 380, 381, 382, 403, 404, 411 **Veselovský, Z.:** 1, 3, 5, 6, 7, 11, 12, 13, 14, 16, 17, 18, 19, 21, 22, 23, 24, 25, 29, 30, 31, 32, 33, 34, 36, 64, 66, 68, 74, 75, 76, 78, 80, 81, 82, 83, 86, 88, 89, 90, 91, 92, 94, 95, 97, 98, 105, 107, 108, 109, 112, 114, 115, 117, 125, 135, 151, 158, 167, 178, 179, 181, 187, 192, 206, 209, 210, 213, 214, 216, 219, 220, 225, 226, 227, 233, 234, 239, 244, 247, 259, 260, 270, 271, 278, 279, 281, 282, 284, 291, 293, 294, 299, 300, 301, 309, 310, 313, 314, 318, 327, 330, 339, 340, 343, 356, 359, 370, 387, 390, 402, 405, 407, 409, 415, 416, 417, 420, 421 **Vergner, J.:** 65, 71 **Vlasák, P.:** 35, 47, 48, 120, 132, 143, 194, 302, 328, 329 **Wormer van, J.** (Coleman): 173 **Ziesler, G.** (Coleman): 57 **Zumr, J.:** 235

Text by Dr Vladimír Hanák
and Dr Vratislav Mazák
Translated by Dana Hábová
Graphic design by František Vlach
Line drawings by Alena Čepická

Published by
CHARTWELL BOOKS, INC.
A Division of BOOK SALES, INC.
110 Enterprise Avenue
Secaucus, New Jersey 07094

Produced by The Promotional
Reprint Company Limited, 1993.

ISBN 1 55521 880 6
Printed in Slovakia by Svornosť, Bratislava
3/11/04/51-03

CONTENTS

Introduction		7
Characteristics of Mammals		8
Adaptations for Locomotion		11
Reproduction		14
Daily and Seasonal Rhythms		17
Territory and Range		19
Distribution		21
Chapter 1	Egg-laying Mammals	25
2	Pouched Mammals	31
3	Insect-eating Mammals	53
4	Flying Mammals	67
5	Man's Closest Relatives	79
6	Anteaters and their Allies	117
7	Rodents	125
8	Carnivorous Mammals of the Land	157
9	Carnivorous Mammals of the Sea	217
10	Horses and their Allies	227
11	Sea-cows	243
12	Hyraxes — a Zoological Puzzle	247
13	Elephants	251
14	Hares and their Allies	257
15	Antelopes and their Allies	261
16	Whales and Dolphins	335
Suggestions for Further Reading		345
Index of Common Names		346
Index of Scientific Names		349

INTRODUCTION

Over the last three decades, mammals have become the subject of unexpected interest, both for the layman and for research workers in various fields. This reflects a general growth of concern for natural life, and the expansion of zoology; another factor is the considerable economic importance of mammals, and not least is the fact that mammals are quite simply a fascinating group, with their many morphological, physiological, and ecological adaptations to a variety of conditions of life and environment. As a single example of the extraordinary variety of mammals, let us consider the striking difference in size of the smallest mammal, the Lesser White-toothed Shrew, weighing slightly more than 2 grams, and the largest mammal of all times, the Blue Whale, sometimes exceeding 150 tons — a weight ratio of 1:6,000,000. The ratio between the smallest and the biggest bird is only 1:45,000. Mammals are remarkably diverse: there exists a wide range of different types within systematically related groups and, on the other hand, mammals similar in shape and ecology are classified into distant evolutionary lines.

Together with birds, mammals rank among the youngest animal classes. Their ancestors appeared on Earth at the turn of the Palaeozoic and Mesozoic eras, 190 to 180 million years ago, as descendants of the inconspicuous, little-specialized reptile order Therapsida. During the long Mesozoic Era, they were quiescent, overwhelmed by the immense expansion of reptiles. It was not until the beginning of the Tertiary that, in a sudden expansion, 35 orders were established; 18 of them have survived, and they represent today the diverse mammalian fauna. In terms of number of species, mammals do not have a leading position. They represent approximately 4,500 species, which is much less than fish (18,000 species), birds (8,600 species), and even reptiles (5,500 species). At present most orders, possibly with the exception of rodents and artiodactyls, have passed the peak of their development. Nevertheless, many of them continue to show vitality.

Of all factors which determined the evolution of mammals, food was the first and foremost. Good adaptative capacities enabled mammals to exploit almost all sources of food available in nature, and accordingly they can be divided into insectivores, carnivores, herbivores, and omnivores. Mammals probably descended from insectivorous, carnivorous, possibly also omnivorous reptiles; this food specialization was also maintained by the mammalian orders which were the earliest to evolve; herbivores appeared later in the course of evolution. The evolutionary influence of food specialization was evident mainly in the composition of the teeth; particularly in the detailed structure of specific types of teeth. These links are so characteristic and obvious that teeth can be used to determine the food requirements of extinct species, and thus provide information about their lifestyles. The study of the effects of food specialization on the composition and structure of dentition is therefore one of the major methods of palaeontological and zoological research on mammals.

This book is too restricted in its extent to permit a more thorough study of the successful evolution of mammals. We will attempt to present a survey of all living mammalian orders, and a comprehensive review of present knowledge on the group. Because of the size of the volume, the most typical and evolutionarily significant species have been selected in describing the respective orders, and the introductory section has concentrated on general features.

We have by no means made full use of the immense quantity of information contained in the wide literature on mammals. The book is intended not only to provide basic facts about mammals, but to stimulate interest in acquiring further knowledge, and above all to point out close links between Man and mammals, so far, unfortunately, exploited by only one of the pair.

Each species is denoted by its scientific name, and the name of the family in which it is classified is also given.

Finally, we would like to raise a question of interest mainly to readers better informed in zoology. It is the problem of domestic animals, their classification and scientific nomenclature. Although domestic animals hardly appear in this book — except forms which could not be excluded in the interest of a complete survey of the individual mammal groups, for example camels, yak, etc. — an observant reader could find certain differences in scientific names if he compared them with other publications. Unlike other authors (mainly German), we are of the opinion that the difference between the present domesticated animals and their not always unequivocally demonstrable wild ancestors, is often much more pronounced than the difference among undoubtedly related species of wild mammals. If we take into account another fact unquestionable today — that modifications of environment, leading to changes in habit, were one of the basic evolutionary stimuli — it seems logical that in the process of formation of animal species (so-called 'speciation'), factors selected in domestication are basically similar to those acting in nature without Man's intervention. In the evolution of Man himself, more precisely in the evolution of the genus *Homo*, the existence of several species is involved, even if the final product of this evolution (that is contemporary Man of the species *Homo sapiens*) is in a way a product of a 'self-domesticating' process. And yet nobody doubts Man's independence as a species! It is a matter of fact that domesticated mammals represent forms with various levels of biological differentiation. In goats and in sheep, for instance, the forms are differentiated on the level of true species (*bona-fide species*), while in the Indian Elephant, influences in domestication were not sufficiently strong to bring about differentiation of a new form at the level of species. Moreover, in one species of domesticated mammals, not all the races (which, to a certain extent, are comparable to individual populations, possibly to geographical forms of mammals living in the wild) necessarily show an identical degree of biological differentiation. In this respect, parallels can also be found in wild mammals.

Now it remains only to thank those who assisted us with preparation of this book, and to express the hope that the following pages will bring readers at least a little enjoyment, satisfaction and learning.

CHARACTERISTICS OF MAMMALS

Mammals are considered to be the most highly evolved class of the phylum of chordates (Chordata), joined by cyclostomes, ancient fish, fish, amphibians, reptiles, and birds in the subphylum of vertebrates (Vertebrata). Mammals therefore have all the characteristics typical of this subphylum, particularly the internal skeleton with an axial spine composed of vertebrae. Many of the progressive adaptations of vertebrates, only hinted at in other classes, have reached their peak in mammals and demonstrate the unique position of this class. It is not easy, however, to present an explicit characterization of mammals, for many of the basic features are possessed in common with reptiles or birds, and furthermore, many typical mammalian features have not developed in all orders; for example, in contemporary monotremes. Members of certain specialized orders (bats, cetaceans, or the bizarre pangolins and armadillos) also differ from the typical mammalian pattern, but the differences concern only the external appearance and way of life. Taking the characteristics one by one, it would be difficult to discover a single one which could absolutely differentiate mammals from other advanced vertebrates. Only the whole complement of qualities, and their mutual harmony and functional inter-relationship, provide the characteristics of the class of mammals. Granting that they are, like birds, direct descendants from reptiles, mammals have become more distinct from reptiles than birds, and they have formed a much more specific group, even though they cannot match birds in terms of number of species.

How can mammals be generally characterized? Most people regard the special way in which they feed the young from mammary glands as the most typical feature. This opinion is reflected in the name of the class. Development of mammary glands is quite unique in vertebrates, although it should be recalled that mammary glands are merely modified sweat glands. Consequently it is

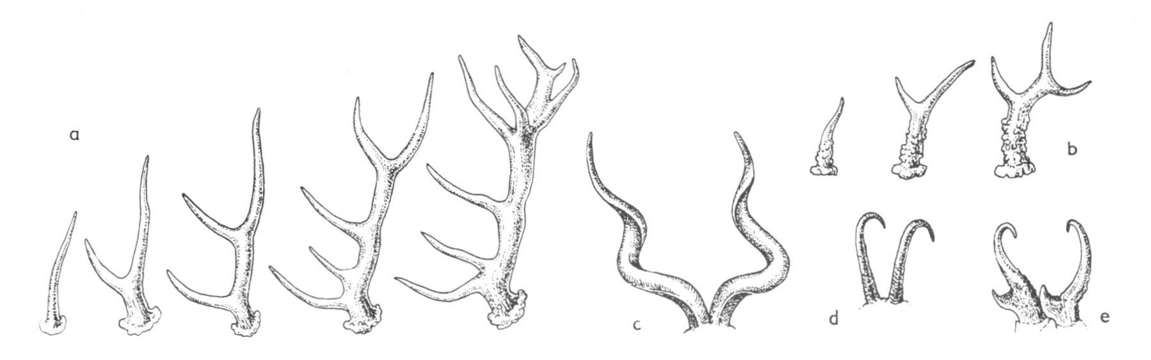

Horns and antlers: a) development of antlers in the Red Deer, b) development of antlers in the Roe Deer, c) horns of the Kudu Antelope, d) horns of the Chamois, e) horns of the Pronghorn

the development of all epidermal glands – mammary, sweat, sebaceous, and scent – that is typical of mammals. Viviparity, and particularly the evolution of the placenta, are also often quoted as characteristic of mammals. This adaptation is, nevertheless, perfectly developed only in placental mammals. The way of reproduction cannot be taken for an entirely unequivocal adaptation of all mammals.

The same applies to another feature which is typical of contemporary mammals, but is not confined to them if we include in the classification extinct groups of vertebrates, – namely the characteristic body cover, hair. Presence of hair has been proven even in pterosaurians (Pterosauria), an extinct group of reptiles. Hair is a horn-like substance growing from the modified epidermal layer of skin, as are the scales in reptiles. Despite that, hair can be regarded as a structure extremely typical of mammals. It consists of two types: the shorter, thick and woolly underfur, and long, straighter and tougher outer hairs. The development of hair is indeed an important mammalian feature, since it occurs in all members without exception; if not in adults, then in the young, or at least briefly during embryonic growth. Accordingly, if in some species or a group (for example in cetaceans, sirenians, elephants, hippopotami) the hair is lacking or reduced, this is always a secondary phenomenon. In other mammals (armadillos and pangolins), the loss of hair went along with the development of a bony carapace in the skin, or of horn-like scales. The loss of hair exceptionally takes place in some underground mammals, for example in the Naked Sand Rat. The spiny coat in hedgehogs or porcupines is nothing but modified hair and claws on toes (present of course also in reptiles and birds) are produced by hardening of the horny skin, as are nails and hoofs. Bovids have other characteristic horny products, horns, made of hollow sheaths set on outgrowths of the frontal bone. Antlers, growing on the foreheads of male and sometimes also female deer, are bony structures; soft skin covers them only at times of growth. Antlers in deer are shed and re-grow annually, while horns in other ruminants are, with some exceptions, permanent organs, growing and expanding with age. The antlers in deer change shape in each generation, mainly in the number of tines.

In relation to hair, we must mention one of the highly progressive adaptation of mammals, which is not unique to them but was of vital importance for their evolution. They are warm-blooded (homoiothermic), capable of thermoregulation, that is of maintaining constant body temperature. Thermoregulation occurs also in birds, and on the basis of the latest palaeontological findings, it probably existed in some specialized Mesozoic reptiles. In present-day mammals, the body temperature is kept at 36 – 40°C; in some marsupials, 27 – 32°C is considered the normal temperature.

Other features characteristic of mammals are related to the structure of the skeleton, particularly of the skull. From a layman's point of view, such differences do not seem important, but they are

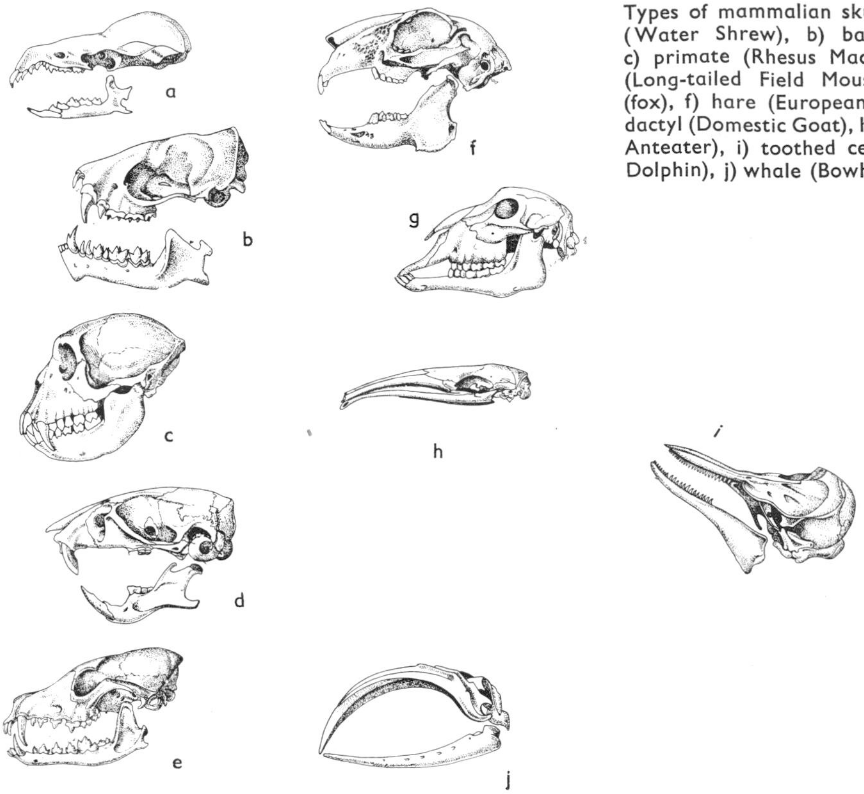

Types of mammalian skull: a) insectivore (Water Shrew), b) bat (Noctule Bat), c) primate (Rhesus Macaque), d) rodent (Long-tailed Field Mouse), e) carnivore (fox), f) hare (European Hare), g) artiodactyl (Domestic Goat), h) edentate (Great Anteater), i) toothed cetacean (Common Dolphin), j) whale (Bowhead Right Whale)

often the only guideline for palaeontologists; they help to separate remains of reptiles from those of primitive mammals. With some exceptions, the skeleton in mammals is completely bony, made of approximately 200 bones. The skull in mammals is the most solid of those of all vertebrates, and, unlike reptiles and birds, it has two occipital condyles – protuberances by which it articulates with the vertebral column. Many bones have entirely disappeared in the course of development of the skull, for example the quadrate, which in reptiles and birds still constitutes an important component in the primary connection between the mandible and the skull. The mandible in mammals is made of a single, paired dental bone (os dentale), joined to the lower part of the skull by a secondary mandibular joint. Teeth in mammals are another typical feature – in spite of a considerable dissimilarity in the individual orders, mammalian teeth differ from those of other vertebrates and advanced reptiles by being divided into four basic types: incisors, canines, premolars, and molars. Mammals are generally characterized by two generations of teeth: simple deciduous or milk teeth, and more complex permanent teeth. Mammals are the only vertebrates with three ear bones (maleus, incus, stapes) instead of one. Primary mammals developed as quadrupeds with five-toed limbs, carrying on from their reptile predecessors. The original five-toed limb underwent a series of changes in the evolution of the individual groups.

Many other characteristic features can be seen in the internal anatomy of mammals. The red blood cells (corpuscles) have a characteristic disc-like shape, and lack a nucleus. The heart is divided into four separate chambers, so that the oxygenated and non-oxygenated blood cannot blend. This, however, is not a unique feature either, for birds and even crocodiles have highly functional hearts of the same structure. Total separation of oxygenated and non-oxygenated blood is in fact an essential adaptation of active animal forms with constant body temperature and more advanced brain function. The mammalian arterial system contains only the left aortic arch.

Peculiarities of other anatomical structures include the muscular diaphragm, which separates the thoracic and abdominal cavities and assists respiration; the elastic cartilage (epiglottis) which closes the larynx; and above all the mobility of the facial muscles, making possible distinct expressions.

The chief prerequisite of successful development of the mammals was probably the high degree of development of their brain, especially of the cortex of neo-pallium, the site of higher nervous activity. The brain of a mammal differs in quality from the brain of any other vertebrate, by its differentiated structure and relative size, although there are manifest differences between brains of primitive insectivores and primates or cetaceans. The extraordinary organization and activity of the brain should therefore be regarded as one of the main and characteristic features of the mammalian class.

ADAPTATIONS FOR LOCOMOTION

The extraordinary variability and adaptability of mammals to different environmental conditions and ways of life are clearly seen in the structure and function of the organs of locomotion. The majority of mammals are quadrupeds with originally five-toed limbs: walking and running are their most typical methods of locomotion. There are pronounced differences in the construction of the limbs of terrestrial forms, affecting their movement and speed of movement. Species requiring speed for hunting prey have reached the highest degree of perfection, and, in contrast, so have the species which defend themselves by fleeing from predators. This applies to many carnivores, the majority of hoofed mammals, hares and kangaroos. The increased speed of locomotion demanded considerable modification of the original type of limb, in particular, elongation of certain structures in the limb, and often a considerable reduction of toes. The limb became straighter, and there was a change in the way of walking, from meeting the ground with the entire lower surface of the foot (plantigrade type) to placement of the flat of the toes (digitigrade type), or only of the tips of the toes (unguligrade type). The Cheetah is considered to be the fastest mammal, capable of pursuing its prey at 90 km/hour. It also possesses the amazing ability to reach the speed

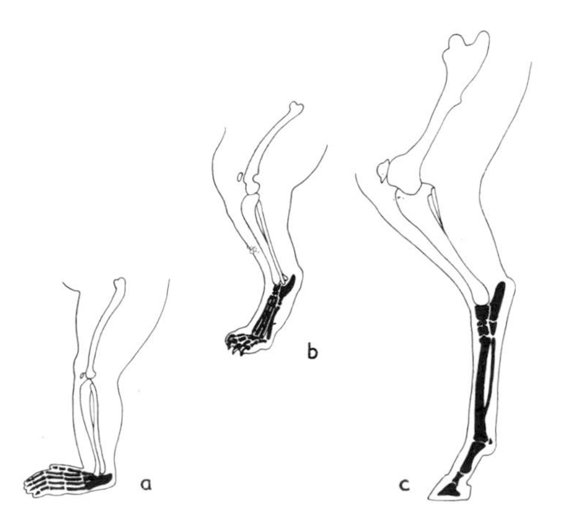

Basic types of limb in mammals: a) plantigrade (monkey), b) digitigrade (wolf), c) unguligrade (horse). Bones of the foot are shaded black

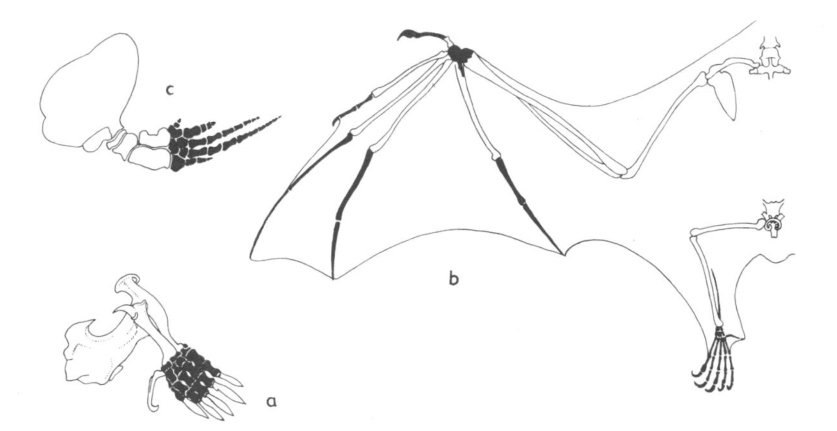

Some outstanding adaptations of mammalian limbs: a) front digging limb of a mole, b) wing of a bat, c) flipper of a cetacean

of 70 km/hour in 2 seconds from standing. Other fast runners are some African antelopes, for example Thomson's and Grant's Gazelles, reaching 75 km/hour; a horse can attain a speed of more than 50 km/hour, and even an apparently clumsy animal like the buffalo can run at 50 km/hour.

The jumping forms of mammals are specialized in respect of locomotion. Progression by means of a series of jumps developed independently in various mammalian groups; in some cases, it led to the origin of forms conspicuously similar in appearance but belonging to completely unrelated groups. In species adapted for jumping the hind limbs are often considerably elongated, whereas the forelimbs become shorter. This is because most of these mammals use predominantly the hind limbs for jumping, and the forelimbs lose their original function in locomotion. However, some species which jump efficiently, for example hares, use all four limbs. Kangaroos are the true jumpers: they propel themselves exclusively by their markedly elongated and powerful hind legs, and the muscular tail aids them in leaping. They can jump as far as 7.5 metres, and they are capable of a speed of 50 km/hour.

The ability to dig in the ground when foraging for food or building shelters is common in many terrestrial forms: many of them divide their activities between time spent under and above the ground. Voles, mice, shrews, pangolins, rabbits, and many others live predominantly in burrows situated underground, but they seek food above. The true underground forms are those which visit the surface only exceptionally, and most of their activity, including the search for food, takes place underground. Among them rank chiefly the moles, golden moles, the strange Marsupial Mole, mole rats, pocket gophers, and sand rats. The various true underground species have many common morphological features, such as a spindle-shaped body, shortened neck and tail, short thick fur without outer hairs, loss of external ears, and often reduction in the size of eyes, which are sometimes completely overgrown by skin. Many underground forms have strong digging forelimbs with long claws, while others have normal limbs and use the powerful incisor teeth (mole rats, pocket gophers).

Numerous mammals found in forests are perfectly adapted for climbing tree trunks and moving in tree tops. This skill is possessed by forest species which are basically ground-living, such as the Pine Marten, Sable, bears, Wild Cat, Bank Vole, and Field Mouse. Squirrels and dormice are even better climbers: they reside predominantly in trees, but move on the ground often and efficiently. The true climbing forms are, however, found among many arboreal primates, sloths, some pangolins, etc. These animals descend to the ground exceptionally and reluctantly; their life cycle takes place in the tree tops. In addition to mobile limbs with sharp, long, arched claws and often with mobile, flexible toes, most species have developed a long tail, which sometimes aids balance in leaping, and in other species serves as a prehensile organ, functioning as a fifth 'limb'.

Many climbing arboreal mammals followed a different line in their evolution; they acquired the capacity of stretching the leaps from one tree to another in a gliding flight. This was made possible by a skin fold on each side of the body between the front and hind limbs and sometimes extended to the neck and tail. The fold is stretched out in flight and functions as a parachute. Gliders are found among rodents and marsupials, but they have reached the highest degree of evolution in the order of flying lemurs (Dermoptera): its two species live concealed in the virgin forests of Indonesia and the Philippines. In these evolutionally ancient mammals (classified close to insectivores, bats and primates), the broad skin fold stretches from the neck to the tips of the toes in all limbs, and to the tip of the tail, facilitating not only longer gliding leaps to a distance of 100–130 metres, but also simple manoeuvring.

The only mammalian order which has fully mastered aerial space, and is capable of active flight like that of birds, is the order of bats. They use the modified forelimb as a flying device: the elongated toes of the limb are covered by a fine, transparent double skin – the flying membrane. The flight of bats can be compared to that of birds, although the movement of the flying membrane is more reminiscent of rowing. Bat's flight is remarkable for its excellent manoeuvrability; in some species, it reaches considerable speed (in the family of free-tailed bats up to 95 km/hour). Many species can glide, but they are incapable of exploiting thermal currents to save energy as do many birds.

When successfully conquering various ecological niches of the world, mammals were not limited by water. Swimming was probably no problem for them – most mammals enter water when it is imperative, and they swim well. Many terrestrial forms move in water with equal dexterity. They are the so-called amphibious species, representing various orders: Platypus, the Otter Shrew, Water Shrew, otters, Nutria, Capybara, Muskrat, Water Vole Rat, and many others. The individual species are adapted to different extents for movement in water. Some have only slight modifications, others are perfectly equipped, for example otters or otter shrews; besides the webbed toes, their bodies are shaped to provide minimal resistance in water: they have a short neck and a thickened base of the tail. Pinnipeds are another evolutionary peak of aquatic mammals: they spend most of their life in water, but the offspring are born on land. Nevertheless, adaptations for aquatic life predominate in them: they have a hydrodynamic body shape, a thick subcutaneous layer of fat, self-closing orifices of the ears and nose, and both pairs of limbs are modified into paddle-like flippers. When swimming, pinnipeds use chiefly the undulating motion of the hind part of the body and limbs.

The true aquatic mammals, never leaving the water environment, are, however, the cetaceans and sirenians. Despite the fact that they are not related, they have developed certain identical adaptations as a result of their similar ways of life. Members of both orders have lost almost all hair, they are spindle-shaped, have a thick layer of subcutaneous fat, and forelimbs modified to flipper-

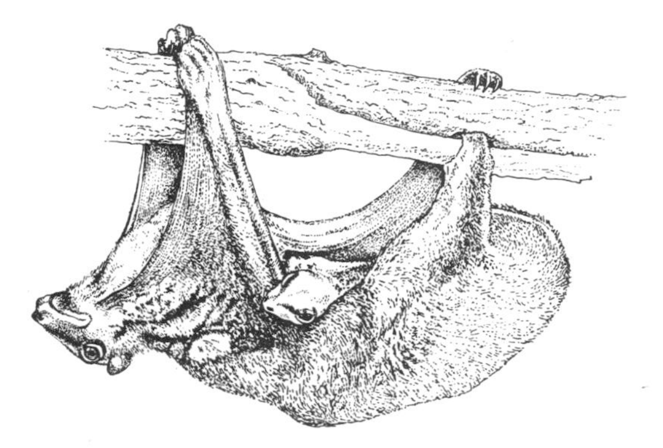

Flying lemur (*Cynocephalus variegatus*)

like organs. Cetaceans resemble fishes in their body shape; they are characterized by the horizontally positioned, forked tail, and a dorsal fin which is usually tall and is made of ligament and fat tissues. In some species the dorsal fin is completely lacking. Other morphological and, especially, physiological adaptations enable them to dive and orientate themselves in water. Thanks to adaptation in an unusual environment, these two orders have developed the most atypical body form of all placental mammals.

REPRODUCTION

Reproduction in mammals has reached the highest perfection; it differs substantially from that in all other vertebrate classes. The majority of mammals are viviparous: the embryonic development takes place within the mother's uterus, which provides both nutrition and defence for the developing embryo. After delivery, the young generally stay under maternal care for a long time; the mother feeds them with milk, defends them from enemies, and rears them. All this considerably increases the chances of survival of the offspring, and their successful introduction into independent life. As a result, mammals are not forced to produce such quantities of eggs as are lower vertebrates, which, by quantity, replace large losses of the young in both embryonic and later stages of development.

Except those of monotremes and marsupials, mammalian eggs are tiny (0.06 – 0.18 mm in diameter), virtually lacking stores of nutritious substances and firm outer layers, and incapable of independent development. The crucial device in the reproductive process of mammals is therefore the placenta, a special organ composed of embryonic membranes (chorion and allantois) and a part of the mother's uterine walls. Protuberances of the placenta, firmly penetrating into the uterine mucous membrane, hold the foetus in the mother's body, transfer nutrients and oxygen from the mother's blood system, remove waste products, and protect the foetus from disease. The true, highly functional, allanto-chorionic placenta is developed only in placental mammals. Marsupials have an imperfect yolk-placenta, insufficient for the needs of the entire foetal development; still in a foetal stage, the young are born underdeveloped, and have to be 'incubated' in the mother's pouch. Monotremes lack the placenta completely: they lay and incubate eggs rich in yolk, in the manner of reptiles or birds.

The allanto-chorionic placenta is a comparatively complex and perfect device, existing in several forms. A relatively simple and free connection with the mother's uterine walls is represented by a *placenta diffusa* in insectivores, prosimians, cetaceans and some ungulates; and by *placenta cotyledonaria* with clusters of chorionic protuberances, developed in edentates and certain ungulates. More solid and naturally more effectively functioning connecting organs are the *placenta zonaria* in carnivores, and particularly the *placenta discoidalis*, occurring in various orders including higher primates and, consequently, Man.

Sexual organs in all mammals are the same in both function and basic structure. In females, they consist of the paired ovaries and Fallopian tubes, with further differentiation into uterus and vagina. Males always have an erectile penis and spermatic cords connecting it with the testicles which are situated either in the abdominal cavity, or periodically or permanently descending to the scrotum. The detailed structure of these organs, however, is distinctly different in monotremes, marsupials and placental mammals, especially in the females. Fertilization is internal in all mammals, and generally occurs in the upper part of the Fallopian tube or in the uterus where spermatozoa penetrate after copulation.

The secondary sexual features are much less obvious in mammals than in other vertebrates. If they occur, they usually involve differences in size of the two sexes, or variations in the distribution or possession of hair, horns or antlers; secondary sexual colouration appears very rarely.

The complex and diverse reproductive processes in mammals depend on the effects of sexual hormones; their production is controlled by the pituitary gland (hypophysis). Activities of the hypophysis are directed by external influences, such as temperature and, chiefly, light — to be precise, differences in the length of daylight which occur throughout the year. Where the seasonal differences in daylight are insignificant (in tropical regions), most mammals are sexually active during the whole year. Such species are known as polyoestrous. The same capacity is found in some species of the temperate zone: they can breed several times from spring to autumn and, under favourable conditions, even in winter. These species are seasonally polyoestrous (many small rodents, shrews, and hares). In many mammals (for example in deer or certain carnivores), oestrus occurs only once a year: these are monoestrous species. Many exceptions and intermediate examples can be found besides these two types. Most domesticated mammals go through two periods of sexual excitement (also called 'heat'): they are dioestrous. Other animals have a 2-year (wolverines, camels, whales), or even 3- to 4-year (elephants) oestrous cycle.

In monoestrous mammals, oestrus occurs in a specific season. In many mammals the period is characterized by specific courtship behaviour, often also by aggressive behaviour of males. Seeking a sexual partner is facilitated by various scent or sound signals, or even by visual signs. Since many species are solitary during the year, the two sexes meet only in the mating period. Males frequently assemble and guard entire harems, and fight for them with rival males. Although in polyoestrous species both sexes encounter more often, they actually live together only in the rutting season. Many mammals live in pairs in the mating season: they are monogamous; others have groups of females around: these are polygamous. The situation in which permanent pairs stay together for several seasons or permanently, as is the case with beavers, jackals, foxes, and certain antelopes, is rather exceptional. Partners are usually exchanged every season, often in the course of one season. Males generally take no part in bringing up offspring.

The period required by the mammalian foetus for its development, that is the time from fertilization to birth, is called gestation or gravidity. The length of gestation is roughly constant for each species: it depends on the size of the species, on its evolutionary perfection, and on the degree of development of the newborn young. Generally speaking, the shortest gestation occurs in the smallest species: in rodents and insectivores. A disproportionately short gestation, sometimes lasting only 8 to 13 days, is characteristic of many marsupials: offspring are born at an early stage of development, and they remain for a long time in the abdominal pouch.

The duration of gestation can differ considerably in species which are closely related or of similar size, for example wild rabbits, giving birth to blind and naked young in sheltered burrows, are pregnant for only 30 days, while hares, dropping furry young capable of vision, have up to 40 days of gestation.

In some mammals, an inordinately long duration of gestation is observed, corresponding neither to their size nor to the degree of development of the newborn offspring. These species undergo so-called 'latent pregnancy', an adaptation in which only the early phase of cell division takes place in the fertilized egg, then embryonic development stops for some time (often several months), and only under more favourable circumstances (usually in spring), is it completed. The aim of this interesting ecological adaptation is to plan the mating, and particularly the birth, for the most advantageous seasons. Another type of latent pregnancy, even better from a biological viewpoint, exists in certain marsupials, rodents and insectivores. Immediately after delivery, renewed copulation takes place, and another generation of eggs is fertilized. Cell division soon ceases for the entire period of suckling. If the first young die, the new embryos start to develop in their stead. An unusual deviation from the usual method of reproduction was also described in true bats of the temperate zone. It is the phenomenon of 'latent fertilization'; following the autumn copulation, sperms remain in a viable state in the vagina or uterus of the female. The following spring, when hibernation ends, the sperms fertilize the ripe egg. Despite the fact that 8 to 9 months elapse between copulation and birth, the true duration of foetal development is approximately 53 to 75 days. In addition, duration of pregnancy in insectivorous bats is influenced by the temperature at the time of foetal development.

The process of birth has been investigated relatively little in wild mammals, since it usually takes place in seclusion, and often at night. Nevertheless, it is known that the majority of mammals give birth in a lying position, some large ungulates and elephants standing, and hippopotami and (naturally) cetaceans in water. Female bats give birth suspended and the newborn bat is caught in the basket-shaped tail membrane. Females of most mammals have to rely on themselves: after giving birth, they bite off the umbilical cord and devour the placenta.

The young of various mammals are born at different stages of development. In ungulates, which deposit offspring in unsheltered sites, the newborn young are well-developed and able to follow the mother without her help very shortly after birth. Carnivores give birth in sheltered locations, sometimes even in burrows: the young are therefore often furry but blind and otherwise little-developed. Pinnipeds or cetaceans drop fully-developed young, reaching generally almost half the length of adults. Most small rodents, insectivores and bats have underdeveloped, blind, naked and feeble young: they must be suckled and nursed for some time in protected shelters, burrows and dens, in the mother's care. Young bats are born naked, blind, with weak wings, but with strong climbing feet and one free toe on the wing.

All this, and special deciduous teeth serving for attachment, enable them in the first days of life to move swiftly on the overhanging walls of caves, and to clutch firmly to the mother's hair.

The rate of growth and development of most mammalian young is extremely rapid in the first phase of post-natal life; it slows down only following the termination of the period of lactation. Many shrews, for instance, double their weight in the first 4 days of life. Newborn mice become hairy in the first week of life, they can soon see, and often become sexually mature before they finish growing. Young bats fly independently from the age of approximately 3 weeks, and very soon become self-reliant. On the other hand, many species, although born well-developed, need a relatively long time, sometimes several years, for further development and particularly for rearing, being dependent on highly developed maternal care. An extreme example are the young of primates, which stay in family packs until sexual maturity, which occurs in some species at the age of ten years. Cubs of many carnivores also stay with the mother for a long time: young wolves until the mother's next oestrus, young Lions for 2 to 3 years, lynxes till the following spring. The long-lasting contact with the mother induces faster development of mental capabilities of the young, in particular perfection of hunting techniques, vitally important for further independent life.

In connection with reproduction, the population dynamics of mammals should be mentioned briefly. The reproductive potential of each species is dependent on an entire complex of various external and internal factors. The most significant ones are litter size, number of births in a year, age at sexual maturity, age composition and health status of the population, and proportion of the sexes. The majority of these factors remain more or less constant, but they can alter to some extent according to environmental conditions. Let us take as an example one of the key indicators of fertility, the litter size. The average number of young in a litter varies in different species and can vary considerably even within one species. There is a rule (with many exceptions of course) that smaller species deposit a larger number of young, and the larger ones have fewer offspring. The greatest number of embryos (up to 32) has been described in Madagascar tenrecs, although it is not certain that all will develop. Domestic mammals have generally much larger litters than their wild ancestors. A small number of young — one or two — is generally found in large mammals, but also in most true bats. The small reproductive capacity in these animals is compensated by a high average life expectancy. Furthermore, the restricted number of young is imperative for them, otherwise the mother could hardly give them the care necessary during the long period of upbringing. Within a species, litter size is greater in northern populations, at times of abundant food supplies and of lower density of population. In contrast, it decreases in females pregnant for the first time, in old females, in times when food is scarce and density of population is too great, and when the health of the population is declining.

Some polyoestrous species, for example small rodents, shrews or hares have an immense potential for reproduction, which nevertheless can be increased periodically under certain favourable conditions. This variation of population density is illustrated by cyclic variations of populations of voles, lemmings, northern hares, and certain carnivores. In Common Voles in Europe and in northern populations of lemmings, such overpopulation occurs at recurrent intervals of 3 to 4 years;

in Blue Hares, Northern Lynxes and foxes the periods are longer, usually about 10 years. The population maximum is naturally preceded by a lengthy period of gradual growth of population density, and it is generally followed by a sudden decrease in numbers, and by a period of low population density. It is now known that overpopulation is a result of periodically recurring times of conditions of favourable climate and food supply: these mobilize all the potential reproductive capacities of the population to a maximum. Litter size reaches its peak, the season of sexual activity is extended, the young mature more rapidly, the sex ratio alters by increasing the numbers of females, and so on. Such staggering expansion has the end result that the density of population surpasses the feeding potential of the habitat, and the population could perish. This does not usually happen; before total disaster sets in, natural mechanisms work in the opposite direction. Food shortage and reduction of the living range in over-multiplied populations cause stress, decline or total discontinuation of reproduction, and diseases. All this, together with increased activity of predators (which have usually also increased in number), causes finally a temporary decrease in density, well below normal.

This illustrates the great flexibility and adaptive nature of the entire reproductive process in mammals. Under normal conditions, fertility of each mammalian species is balanced in order to secure continual replacement of population in the density which is best for the species in the given circumstances. Reserves are always available, enabling the species to adapt to modified conditions and replace unexpected losses, and assisting its further development and distribution.

DAILY AND SEASONAL RHYTHMS

More than a half of all mammals are nocturnal — that is they move about, seek food, clean themselves, build burrows and shelters, reproduce, mark and defend their territories at night. They take advantage of the protective darkness which screens them from enemies, and they are better able to exploit their most sensitive sense — smell. Nocturnal mammals comprise chiefly small species, for example rodents, insectivores, hares, most endangered by various predators. All bats are a specially adapted nocturnal group, and so, naturally, are the majority of predators of small mammals. Many of them start their night rambles and hunts in complete darkness, while others reach the peak of their activity at dusk and dawn; these are called 'crepuscular' animals. Most crepuscular species leave their shelters twice a night, and rest in between. This is the so-called two-phased type of activity, unlike the one-phased activity of animals which set out to hunt in the evening and return in the morning. There are few true diurnal types among mammals, but they include, for example, squirrels, hoofed mammals, and all higher primates including Man. They are mostly animals which do not have to fear enemies, since they can either flee or they are sufficiently big and powerful to oppose them. Sight is the principal organ of diurnal mammals, but hearing is also well developed. In many species, activity is not accurately synchronized with alternations of light and darkness, and takes place during both night and day. This happens, for instance, in shrews and many small rodents which are forced to supplement their food stores several times a day because of their rapid metabolism. Their activity is consequently divided into a larger number of short periods; this is known as poly-phased type of activity.

In reality, this categorization of mammals according to the type of activity is highly unreliable; the situation in the wild is much more complex. The basic rhythm of activity can vary considerably in individuals of the same species. For instance in circumpolar regions, where the regular rotation of day and night does not occur, many species become poly-phased, and they alter their activities with respect to their needs. Synantropic species, for example Norwegian Rats and other rats, are equally noted for fluctuating and adaptive activity. They match their behaviour to the working schedule of people in the vicinity: in agricultural enterprises, they become active after the termination of the working day or at the time when domestic animals receive food. Seasonal changes in the rhythm of activity are also common. Many shrews and small rodents are active predominantly at night in the summer season, and when temperature decreases in autumn, they transfer their activity to the daytime, when it is warmer. Duration and division of activity usually change with increased density of population, in periods of starvation, in unfavourable weather, and chiefly in the mating season and during rearing the young.

The rhythm of activity of each animal species is controlled by a set of external and internal signals and regulators, essentially by alternations of light and darkness, by temperature, food supply, and by the inherited daily programme of activity, which is called the internal or biological clock.

Similar to the daily rhythm of activity is the cycle of mammalian activity during the year. This is most noticeable in mammals living in the temperate zone, where the environment in winter is drastically changed, food becomes scarce, and demands on the thermoregulating capacities increase, etc. At that time many mammals change their feeding habits, switching over to alternative nutriments, or living off stores gathered in autumn. Many small mammals opt for life under the snow cover when winter sets in. This is advantageous, as they are protected against severe weather conditions (even at −50°C air temperature, snow temperature does not decrease to less than −5°C), against many predators, and furthermore, they have stores of green food preserved by frost. The mobile, large species migrate before winter to more suitable sites.

The most interesting seasonal adaptation of some mammals is their capacity of surviving the adverse season in the so-called winter sleep or hibernation. It is a state of torpor, very similar to normal sleep; all the body functions are reduced to a minimum in order to diminish energy consumption, and to avoid the need for food. True hibernation therefore developed in the species of the temperate zone, feeding chiefly on insects or green parts of plants — in hedgehogs, bats, dormice, sousliks, hamsters, etc. The ability to halt the body's thermoregulating mechanisms temporarily is, however, quite common in mammals: these tendencies are found in some inhabitants of warm areas, especially in the evolutionally ancient egg-laying mammals and some marsupials, but also in prosimians and edentates. Many other inhabitants of more northerly regions of the temperate zone undergo so-called 'false' hibernation, when the body temperature, heart rate, and rate of metabolism change only negligibly. Such a dormant state can be observed in bears, badgers, raccoons, and raccoon dogs; the animals often emerge, sometimes for a considerable time, and they take in food. Females of bears even give birth to their young and suckle them.

The major stimulus for starting hibernation, although not the only one, is the autumnal decrease in temperature. It is enhanced by the shortening of the day, lack of food, fattening of the animal after autumn foraging, occupation of an adequate shelter, and even augmentation of the amount of carbon dioxide in the unaerated environment of underground dens. The changes in the environment are detected by the brain, which then produces stimuli limiting the activity of certain internally-secreting glands.

Most hibernating animals spend winter underground in dens lined with dry grass. They are generally totally immobile, in the characteristic dormant position. Dormice, for example, are rolled into a ball, heads covered by bushy tails; bats creep into cracks in walls and ceilings of underground caves, or hang from ceilings of houses. When touching a hibernating specimen, it can be felt how cold it is; its body temperature does not differ much from the temperature of the environment. In some bats, the body temperature can decrease below zero for a limited time; most hibernating animals, however, wake up under such circumstances, and move to a more suitable place. Respiratory and heart activities are also substantially slowed down. In sousliks, which take 150—200 breaths per minute in the active state, the number of breaths decreases to 1—4 in a minute. In bats, 3- to 8-minute intervals elapse without breathing, followed by 3-minute periods with 25—50 breaths per minute. In rodents, the number of heart beats is reduced from 200 to 5—18 per minute. In most cases, hibernating animals meet the needs of their reduced metabolism merely from the stores of body fat, made up in late summer or in autumn. Hibernating animals generally do not respond to any external stimuli except temperature, which testifies to the fact that their sensory perception and brain activity are suppressed. Yet it would be wrong to imagine that animals might spend the entire winter season — often almost as much as 7 months — in this state. It has been shown that, from time to time, they emerge for a shorter or longer time, fulfil their physiological needs, sometimes eat a little, stretch their limbs, and continue the interrupted hibernation. Hedgehogs and hamsters wake up quite often, bats less frequently and sousliks, dormice, and marmots very rarely.

Some semi-desert and steppe mammals, for example sousliks, jirds and jerboas, pass the most oppressive summer heats, droughts, and periods of green food shortage in summer sleep or aestivation. From a physiological viewpoint, this state does not differ from hibernation. Aestivation sometimes overlaps uninterrupted into hibernation, and the animal thus spends most of the year resting.

TERRITORY AND RANGE

Most people believe that animals living in the wild are absolutely free, moving unrestrained, and caring only about food and protection against enemies. Such supposition is, however, quite wrong. As far as mammals are concerned, not only is each species closely bound to a specific environment, for example a deciduous forest, steppe, or prairie; but, in addition, each individual, pair, or a larger social group inhabits a certain limited area called the home range, or territory. The two terms are distinct in meaning, but for the purposes of this book, it is sufficient to understand that the term 'territory' relates to an area actively defended by the animal and situated within a somewhat larger 'home range'. All activities of the individual or group take place here; they seek food, move around, rest, mate, bear and rear their young. Every owner knows its territory in detail, knows about the richest sources of food and the shortest paths, and protects its territory from enemies and, naturally, from other members of its own species. Many mammals, for example voles, partially improve their territories or home ranges: they construct pathways to sources of food, concealed food stores, and escape routes.

Since the majority of mammals are macrosmatic (with keen perception of smell), they mark the borders of home ranges chiefly by odours — exudates of scent glands, sometimes also excrement or urine. The signs immediately warn every intruder that the area is occupied, and in addition provide precise information on the sexual activity of the owner, of its momentary mood, possibly of the time or duration of its stay. Some mammal species mark their territories with visual signs: bears, for example, leave traces of claws on tree bark. Others announce their presence and ownership by specific vocal performances, for example many primates, Lions and canids.

Territories differ in size according to the size of the animals, their food demands, and other factors. Generally speaking, a territory has to be sufficiently extensive to secure ample food for all its inhabitants. Voles, consuming exclusively green food, are content with space of several dozens or hundreds of square metres. Field mice, feeding on seeds of certain plants, and particularly insectivorous shrews, have much larger territories, in dimensions of thousands of square metres. Large mammals have territories corresponding to their size; but they are always smaller for herbivorous ungulates than for large beasts of prey. For instance, the Brown Bear hunts in a territory of several square kilometres; Tigers or wolf packs have even more vast hunting ranges. The size of a territory can differ considerably within one species. This depends on the geography of the area and the abundance of food, on the season, and on population density. It was discovered that females and juveniles have more restricted territories than old and sexually active males.

Some non-territorial species show interesting patterns of behaviour. It seems that cetaceans always follow the same routes, but they probably have no home ranges. Pinnipeds also confine territorial behaviour to the mating season only, when they struggle for smallish coastal ranges for their harems of females and young. In the reproductive season breeding territories are protected also by many herd-forming ungulates of the African savannas, for example by gnus.

Although the instinct of the home range is well developed in most mammals, it cannot be said that they are very active in building shelters or nests. Underground burrows are the most frequent type of actively built shelter: they are dug by members of various orders. Individual types of underground shelter nevertheless differ widely in their basic construction, depending on whether they serve as temporary retreats or permanent homes, or nests for offspring. The true underground mammals — moles, mole rats and pocket gophers — build the most elaborate networks of underground burrows, nests and food stores. Underground refuges of rodents living in colonies, for example prairie dogs and marmots, are also highly intricate; they are usually inhabited for a long time, constantly enlarged and repaired. Characteristic lairs with dens are dug out by certain beasts of prey. Members of less common orders, for example pangolins and aardvarks, also hide and rest in burrows. The Otter digs its burrows in banks of rivers or stagnant waters; also, under certain circumstances, muskrats build up piles of roots and stems of aquatic plants (muskrat lodges) in shallow water, with a chamber for offspring in the middle above water level. The advantage of these constructions lies in the fact that entrances do not freeze in winter, and muskrats get under the ice easily. The Beaver is one of the best builders among mammals: besides its lodges, it also erects dams, regulating the water level within its territory. Elaborately woven nests of stems, hung on plants or bushes, are fabricated by some mouse and dormouse species. Many mammals settle in various suitable shelters — for example tree hollows, under uprooted trees, rocky crevices, small caves, and often sections of human dwellings — where they supply only the lining. Some species are fond of occupying finished homes of other mammals, or of animals from completely different groups (termitaries, birds' nests, etc).

Many mammals are content with any shelter they discover when rambling through the territory. They never furnish such random retreats, and soon replace them by others. This is typical of many ungulates. Monkeys do not build special shelters either, although one might expect it in view of their advanced intelligence. Apes are known to rig only simple refuges of branches in or under trees, usually new ones every day, and anywhere where they are before nightfall.

So far, we have mentioned only mammals which could, from our point of view, be called settled; but there are also mammals which are true nomads: they do not move only within their territory, but for certain reasons set out on long journeys. The regular seasonal migrations, comparable to passages of birds, occur rarely in mammals, and they are usually motivated by a search for suitable food sources, more favourable climatic conditions, shelter, or by sexual activity. Northern reindeer regularly move southwards before winter, and return in spring. Thus, they can spend winter in places with less snow, which allows them to reach food. Likewise, herds of African ungulates, zebra, gnu and antelope, move in cycles throughout the steppe, gradually grazing up the grass. Regular migration of some pinnipeds, particularly sea lions, from the open sea to certain islands, is in harmony with their reproductive cycle. Many bats of the temperate zone also rank among migrating species. The reason for their annual passages from summer sites to winter locations and back is provoked not so much by food requirements, but by the search for suitable shelters for hibernation. Flights of most species are therefore short, in the range of several dozen kilometres, 100—200 km at the most. Some species, on the other hand, effectuate really long-distance passages, comparable to those of birds. In North America, they include the genera *Lasiurus* and *Lasiopterus*, in Eurasia genera *Nyctalus*, *Pipistrellus* and *Vespertilio*. Noctule Bats fly every year from northern parts of their European range, overcoming distances of thousands of kilometres, to hibernate in the milder climate of France, the Balkans, the Crimea and Caucasian foothills. Some large cetaceans also accomplish long-distance migrations as the mass distribution of their food, the sea plankton, alters during the year according to the sea temperature. It seems that each species of migratory cetacean has its fixed routes. Some whales can swim annually a distance of more than 10,000 km.

Certain mountain ungulates, for example ibexes, wild sheep, Chamois, and mountain populations of deer undertake shorter seasonal migrations. In winter, they gather in valleys and forest zones, where they find better climatic conditions and more food. Even some small rodents, belonging for most of the year to the settled types, become more mobile in autumn. The forest and field types of shrew and field mouse find winter retreats in dwellings, often several kilometres away.

Non-periodic migrations of some mammals take place regardless of seasonal cycles, under the influence of various factors, usually following an increase in population density. An example of such migrations is the migration of lemmings, northern rodents, which is triggered off by cyclic overpopulation in certain parts of their territory. It is in fact a total mass exodus from the overpopulated area, with indications of stress and aggressive behaviour. The lemmings always go in a single direction, and they never settle in another site, for the migrating populations are gradually eliminated by obstacles, hardship, and by predators accompanying the passage, and probably by other internal causes also. Migrations of lemmings are, then, an extreme case of emigration of young components of the population, occurring less conspicuously in many other species with a rapid reproductive capacity. The surplus young of all territorial species have to emigrate into other locations when they have become independent, and to find their own home ranges. Emigration is a necessary prerequisite of spreading and enlarging the area of distribution of each species. Distribution of muskrats is a classic example: ten years after their introduction, they spread over all of central Europe. At present, Raccoon Dogs, Moose, and other species of game mammals are expanding through Europe; thanks to protection and a reduction in the numbers of their predators, they can exploit their surplus production.

DISTRIBUTION

Mammals have certain advantageous adaptations: they are homoiothermic and viviparous, and, thanks to these qualities, they colonized all continents, and cetaceans inhabited all oceans and inter-connecting seas. The most variegated and rich communities of mammals, in terms of species, were formed in tropical regions and subtropical open landscapes, but they can be found even in the extreme conditions of deserts, high mountains and the northernmost areas. In the Arctic regions, many species reach the northern edges of continents and certain resistant forms such as the Arctic Fox and Polar Bear extend over the ice to even more northerly situated islands. In the Southern Hemisphere, mammals other than pinnipeds do not cross the southernmost borders of the continent, but several species of rodents, two carnivorous species, and one species each of bats and llamas occur as far south as Tierra del Fuego. The majority of oceanic islands, often thousands of kilometres from the nearest continent, have their, often very restricted, mammalian fauna composed of bats and rodents.

Some mammalian groups or species have immense areas of distribution, often extending beyond continental borders. For example, true bats and free-tailed bats are distributed throughout the world. Similarly, all continents except Australia are the homes of families of hares, hamsters, squirrels, canids, weasels and felids. At the opposite extreme, there are species with a very small range of distribution, one of them being the Volcano Rabbit (*Romerolagus diazi*), confined just to a particular altitude on slopes of the two Mexican volcanoes, Popocatepetl and Iztaccihuatl.

The basic map of the contemporary distribution of terrestrial mammals corresponds to the division of the world into its main zoogeographic regions, roughly demarcated by zoologists in the 19th century. Each of these five vast areas is characterized by a typical mammalian fauna, which colonized them during a long, often highly complicated history. Major factors in the development of the fauna in these areas were the distribution of original centres of individual groups, their ecological adaptability, and chiefly the effectiveness of the so-called barriers limiting the spread of faunas and in many cases designating the boundaries of their ranges. The principal barriers were the sea, sometimes even narrow straits, large rivers, mountain ranges, and deserts. The influence of the barrier frequently remains permanently, even if the obstacle disappeared a long time ago, as in the case of the specific fauna of South America.

The Australian region is the home of the most remarkable and best differentiated mammalian fauna in the world. It is dominated by numerous forms of marsupials, and even a few forms of monotremes have survived there, existing nowhere else. They are undoubtedly the remains of a very ancient fauna which survived here and, because of the early and long-lasting separation of the Australian continent from the rest of the world, developed successfully without competitive pressure from the placental mammals which evolved later. Yet, not even such a perfectly separated area was forever closed to the expansion of mammals of more recent derivation. It was discovered by the mobile bats and even by mice, which spread gradually, in 'jumps', from one island to the next. Nevertheless, the primitive character of the Australian fauna has been preserved and allows us a glimpse of the mammalian life in the bygone Tertiary period. The boundary between the Australian and Oriental mammalian faunas in the region of the Sunda Islands, can be traced along the so-called Wallace line between the islands of Bali and Lombok, and between Borneo and Celebes.

The Neo-tropical fauna of South America has undergone evolution somewhat similar to that of the Australian fauna. The species composition differs in the two areas, but they have in common a distinct individuality, which, in the case of Neo-tropical mammals, can be explained by the long isolation of the South American continent in the Tertiary. The most ancient components of the South American fauna, the marsupials, some extinct primitive ungulates and the ancestors of the edentates, settled here at the very end of the Mesozoic and in the early Tertiary, when the connection with North America still existed. Throughout the Tertiary, these groups formed a dominant and varied mammalian community. In the course of the Tertiary, however, before the re-formation of the land bridge between the two Americas, ancestors of the New World monkeys and particularly of caviomorph rodents had arrived by 'island hopping' from North America to South America. Today they constitute a vital and important component of the mammalian fauna. From the end of the Tertiary and throughout the Quaternary, invasions by other mammalian groups, notably carnivores, brought about the extinction of most of the original South American mammals. The extinction included most of the ungulates, carnivorous marsupials and edentates. This resulted in radical 'modernization' of the mammalian fauna. The process of mutual exchange of faunas between North and South America is extremely interesting and instructive. While the invasion of 'modern' mammals from the north southwards was tumultuous and extensive, the advance of the original South American mammals northwards was considerably restricted; only some edentates and marsupials succeeded, although many soon became extinct.

The Holarctic region is the largest zoogeographic area of the world. It includes a major part of Eurasia, northern Africa and North America. Since most of this area is situated in the temperate zone, noticeably affected by glacial periods in the recent past, the Holarctic fauna is relatively poor in number of species in comparison with that of tropical regions. Its typical mammalian representatives are the deer family, squirrels and beavers of the rodent family, and moles in the insectivores. Most of these families penetrate into other areas also, particularly to the Ethiopian and Oriental regions. There is an interesting relation between the American (Nearctic) and the Eurasian (Palaearctic) parts of the Holarctic region. Many species occur in both sub-regions, and many more groups are represented in the two sections by different, but closely allied species. The most outstanding similarities between the two subregions exist in the mammalian fauna of tundra and taiga, where spread in both directions was facilitated by the Bering Strait. Examples are the circumpolar distribution of Reindeer, Polar Bears, lynxes, lemmings, Moose, Beavers, etc. In the community of steppe and desert mammals, the differences between the two sub-regions are much more striking. In America, families of mice and jerboas, so typical of the southern parts of the Old World, are totally absent; many rodents, ecologically replacing mice and jerboas in America, are lacking in Eurasia. Both sub-regions have their native families — for instance Pronghorn Antelopes in America and mole rats and selevinias in Eurasia.

Although the mammalian fauna in the two last large regions, Ethiopian and Oriental, appears at first sight to be different from that of the other regions, and highly specific, from the purely zoological viewpoint it demonstrates very close relationships with the Holarctic fauna. The most conspicuous differences result from the location of these two areas in the tropical zone; they have therefore a highly variegated and rich mammalian fauna, which was not so drastically reduced by the climatic changes in the early Quarternary. There are, however, no insurmountable barriers between the three zoogeographic regions of the Old World, and the majority of mammalian families are common to all three (or they were, in the recent past). This is not to imply that the Ethiopian area, for example, does not have its characteristic mammalian fauna. It is inhabited by otter shrews, golden moles, elephant shrews, and even one exclusive order, the aardvarks. The impression of Africa is, for many people, one of large herds of various ungulates and beasts of prey populating the savannas. These are, zoologically, relatively young settlers, yet there is no doubt that these groups went through a turbulent stage of development on the African continent, and created an imposing mammalian community, unlike that anywhere else in the world. Other typical African mammals, such as various groups of primates, rhinoceroses, elephants and pangolins, are nowadays resident even in the Oriental region, demonstrating the close relation of the two areas, and a similar historical development of the two faunas. Naturally, the Oriental region has its own native mammalian residents: tree shrews, flying lemurs, tarsiers, etc.

The island of Madagascar is a singular zoogeographic refuge; although it is regarded as a part of the Ethiopian region, it is the home of an entirely specific mammalian fauna, which became extinct in the African continent a long time ago. It is inhabited by the peculiar prosimians — lemurs, in a variety of forms — by the primitive insectivores (tenrecs), by Madagascar hamsters, and by several species of carnivore.

In conclusion, it should be mentioned that the processes of distribution of mammals over the world were often extremely intricate, and that we cannot always deduce precisely the origin of individual groups or species on the basis of their present distribution. For instance, marsupials are not originally Australian, although they now represent the dominant Australian group; originally they were widespread in at least four present-day zoogeographic regions. From fossil remains, we know that tapirs, for example, inhabited North America and Eurasia in the Tertiary Era. Their present isolated distribution in the Oriental and Neotropical regions are again only the remains of the original wide range. Ancestors of elephants originated on the African continent, but they later moved to Eurasia and North America, where the essential stage of their further development took place. Their recent ancestors then returned to the aboriginal lands, which, with the near Oriental region, have today become their only refuges. A similarly complicated tour of the world was undertaken by horses, camels, rhinoceroses, and probably many other mammals also.

Nevertheless, not even the present distribution of mammals is completely permanent and unchangeable. Immense changes in the mammalian fauna of the Holarctic region are connected with the geologically relatively recent glacial periods and the individual climatic post-glacial periods. The present distribution of mammals is subject to many relatively radical changes, mostly caused by Man's activities. Owing to Man, many species of mammals became extinct in the last century, and others were forced to decrease their range of distribution. On the other hand, many others have been assisted by Man, directly or indirectly, to a vast distribution.

1

Chapter 1 EGG-LAYING MAMMALS

When the first platypus and echidna were brought from Australia at the turn of the 18th and 19th centuries, their appearance amazed even the most renowned zoologists, and it took them many years to agree that these creatures really were of the mammalian class. All the confusion and heated scientific disputes were provoked not only by the looks of the weird Australian species but mostly by their internal anatomy, and by the fact that they were laying and incubating eggs, like reptiles or birds.

It is no mystery today that the monotreme order (Monotremata), as these mammals were named, represents the last surviving remnants of prehistoric mammalian types, and brings together a complicated mosaic of both advanced mammalian features and primitive reptilian ones. It is their restricted distribution, unhindered by competition from more progressive forms, and their extreme specialization that allows them to survive. Monotremes probably developed independently from Mesozoic mammal-like reptiles (Therapsids) or from primitive mammalian lines off the dominant trends in the mammalian evolution. They are therefore included in the special subclass of egg-laying mammals (Prototheria).

They are the only mammals to possess a cloaca — the common cavity into which the intestinal, urinary and genital tracts open. Other reptile features include: the internal arrangement of the urino-genital system, the eye, the inner ear, some bones in the skull, and the shoulder-blade and pelvic areas. Their eggs are large, rich in yolk and covered by a skin-thick coating. Embryos develop during incubation under the mother's warm body, but they soon become typical mammals: they are furry, warm-blooded and when young, feed on the mother's milk.

Australian Echidna
Tachyglossus aculeatus
Tachyglossidae

(1, 2) is about 45 cm long and weighs 2.5—6 kg. The back and flanks are covered by spines, 6—8 cm long, which are erected in case of danger, when the animal rolls into a ball. The sparse hair is almost imperceptible among the spines and is only noticeable on the abdomen. The Echidna's

2

3

head is relatively small, with beadlike eyes and external ears without lobes. The slender tube-like snout is about 8 cm long, the mouth is smallish and the nasal apertures tiny. The legs are rather short, but strong and muscular, with long and flat claws: the conspicuously elongated claws of the second toes on the hind legs serve as spine-cleaners. Although the Echidna is one of the most ancient and primitive living mammals, it is a very successful and widely distributed species. It is widespread in Australia and New Guinea, residing mostly on rocky grassy slopes, but also in woods and sandy flatlands from lowlands up to 2,500 metres of altitude. As it lives a secretive and nocturnal life, active in the evening and at night, it can rarely be seen; during the day it hides in rocky crevices, hollow trees, or in burrows which it has made. Its diet consists chiefly of ants, termites and other invertebrates, which it digs out, sticks on its extensible tongue and crushes with the horny palate and hard stomach lining. Its body temperature ranges between 19° and 32°C according to the temperature of the environment. It is quite resistant to the cold and in the southern section of its territory it even hibernates. On the other hand, the Echidna is extremely sensitive to higher temperatures and perishes if its body temperature increases to more than 38°C. The Echidna's breeding habits were described almost a hundred years after its discovery: it lays one, exceptionally two or three eggs, 13–15 mm in diameter, enclosed in an elastic leathery membrane. The mother places the eggs in a deep pouch on her abdomen which develops exclusively in the mating period, and the eggs are incubated there for 7–10 days. Afterwards, the young, which measures merely 12 mm, cracks the egg shell with the sharp egg-tooth on the tip of its snout and crawls into the pouch, where it sucks the thick milk exuded from milk patches on the mother's abdomen. The young stay in the pouch for 6–10 weeks, growing to a length of 10 cm; the mother then hides them in a sheltered place and continues to suckle them for quite a long time. Young echidnas reach sexual maturity at the age of one year.

Bruijn's Echidna (3, 4)
Zaglossus bruijni
Tachyglossidae

is one of the three large echidnas found in New Guinea. It has long legs, a long snout slightly bent downwards, conspicuous ear lobes and short, widely spaced spines. It measures up to 80 cm and weighs up to 10 kg. It was discovered late, at the end of the 19th century, and its way of life has so far remained little known.

4

Young Platypuses lick milk from milk patches on the mother's abdomen

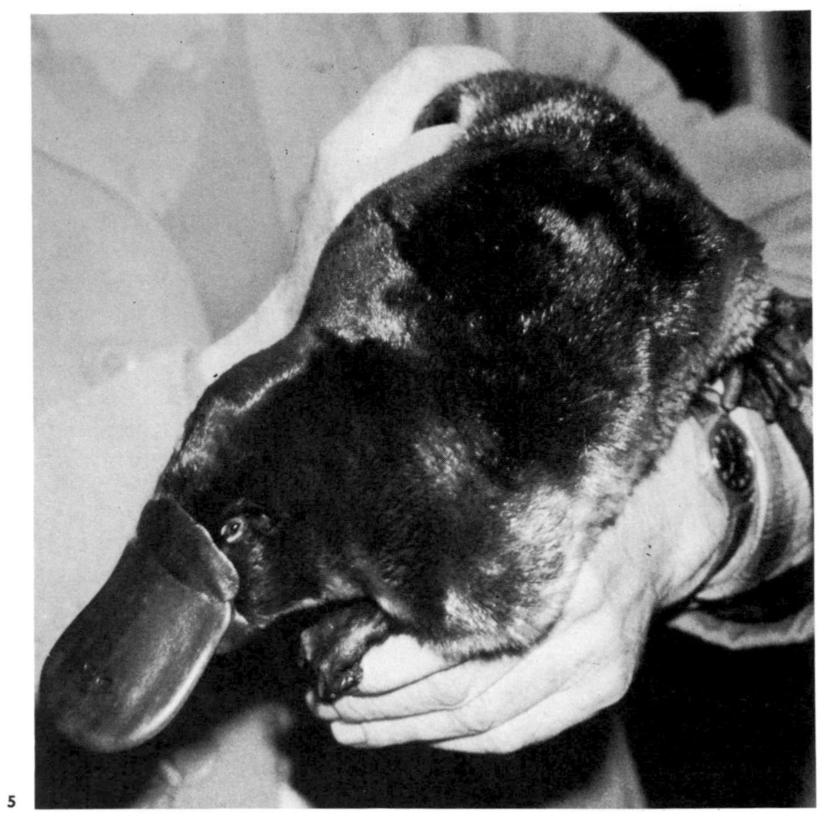

5

Duck-billed Platypus
Ornithorhynchus anatinus
Ornithorhynchidae

(5, 6) is confined in distribution to the south-eastern part of Australia and to Tasmania, but it is not so widespread as the echidnas. It measures about 50 cm (including the long, flattened tail), and it weighs only 0.5–2 kg. The cylindrical body is completely covered by thick fur composed of coarse guard hairs and soft silvery undercoat. The stout, flattened feet bear five toes with sharp claws connected by a tough swimming web. The Platypus is restricted to banks of still waters and rivers up to 1,600–1,700 metres of altitude: its body is well-adapted to swimming and diving. The Platypus spends most of its active life in water, where it obtains most of its food: small crayfish, worms, aquatic insects and molluscs. The 'bill' of the Platypus is an ideal tool for catching and filtering out the tiny prey. Its jaws are furnished only with flat, horny plates; real teeth appear only temporarily, in the young. The food is gathered mostly in the evenings and early mornings, is stored by the Platypus in its facial pouches and consumed on land. The quantity of food consumed per day corresponds to almost half of its body weight.

Platypuses live individually or in pairs. They dig burrows in banks, where the female lays usually two tiny eggs, which often get glued together by their soft membranes. The female curls around the eggs and warms them for about 10—12 days. She rarely leaves the nesting chamber and refuses food during this time. The young are born naked and blind, about 25 mm long, and they suck milk from primitive milk glands on the mother's belly. After another four months they are able to move independently outside the nest and to find extra food for themselves. These interesting observations on the reproduction of Platypuses were gathered during World War II in a southern Australian preserve, although the Platypus had been described already in 1799. This delay was certainly caused by the fact that it is very difficult to keep platypuses in captivity, and even today they can only rarely be found in zoological gardens outside Australia. In Australia, the Platypus is strictly protected within wildlife reserves. Although fully protected and not pursued for its precious fur any more, it is still menaced by fishermen and trout breeders who blame the Platypus for devouring fish roe.

Finally, an error should be corrected which has been embodied in both the Latin and many foreign names of the 'duckbill'. The Duck-billed Platypus, despite a remarkable resemblance of its jaws to a duck's bill, has nothing in common with birds. Some of its unusual features prove its relation to reptile ancestors, and most of them are identical with the peculiar characteristics of echidnas.

6

7

Chapter 2 POUCHED MAMMALS

No mammalian order can offer such an extensive morphological variety as the order of marsupials (Marsupialia). Among the 250 species so far known, there are forms similar in appearance and way of life to moles, mice, shrews, monkeys, dormice, various beasts of prey and even ungulates. It is, then, reasonable to ask 'where did such a variety come from?' — for it seems as if, within one order, the 'inventions' of other orders of placental mammals had been imitated or almost duplicated. There is only one explanation: all these forms have gradually developed from smallish predacious ancestors externally similar to mice, by the process of the so-called 'adaptive radiation', that is adaptation to conditions of life in different environments. Such evolution is only possible in isolated areas, without threat from more advanced forms. Despite the variety in shape, marsupials share many typical, common features, differentiating them from other mammals: a relatively small cranium, primitive brain-structure, certain differences in the skull bones and in the number of teeth. The abdominal pouch (marsupium) is of course the most popular marsupial feature: it is supported by special protrusions of the pubis, the 'pouch bones' (ossa marsupialia). The pouch shelters milk nipples, and the young are reared there. It is, however, only one of the peculiarities of this unorthodox means of reproduction: almost all marsupials lack a real placenta, and during their brief uterine development the embryos are fed by the uterine mucous membrane through the walls of the yolk sac. The young are born as tiny, helpless creatures, which really 'hatch out' only after a relatively long stay in the pouch.

Marsupials are a very ancient group of mammals which originated at the end of the Mesozoic and at first colonized almost all the prehistorical world. Nowadays, they are confined to Australia and South America, with only a few species being found in North America.

Mouse Opossum (8)
Marmosa murina
Didelphidae

is one of the representatives of about 50 species of small, mostly arboreal forms, distributed from central Mexico to Patagonia. The pouch is not developed; the young suck from nipples arranged in two rows or in a circle on the abdomen.

8

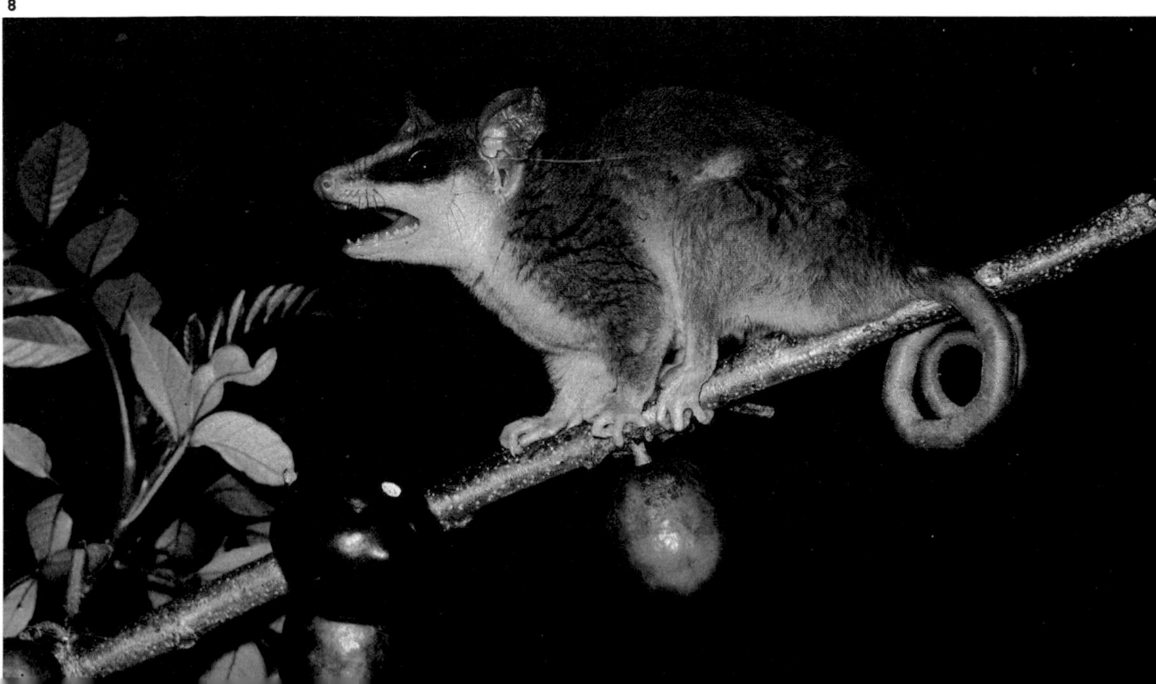

North American Opossum (9) is distributed from South America to most of the USA and southern Canada. It resembles a large rat, measures about 40 cm and weighs 2–5 kg. The fur is thick and greyish, the tail bare and prehensile. The female's pouch opens forward and the milk nipples (mammae – up to 17) form a circle. Usually one litter is produced each year in the north, and two or three in the south, averaging eight to 14 young, exceptionally as many as 25. The little developed, bean-sized young are born after 12–13 days of gestation and remain in the pouch some 60–70 days, firmly attached to the nipples. For another month after that, they stay on the mother's body, clutching to her back and coiling their tiny tails around hers. They feed on her milk and only at the end of this period start to consume solid food. The North American Opossum lives in forests and bush and near human settlements. It builds nests of leaves and grass in rocky crevices, hollow trees or in other mammals' burrows, sometimes even in buildings. It is active at night, and is mostly directed

Didelphis marsupialis
Didelphidae

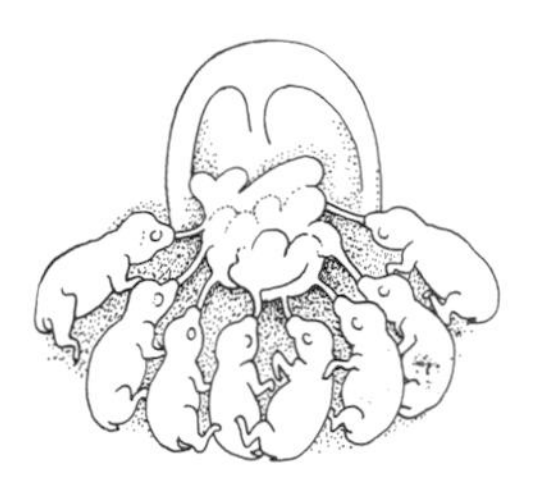

Young oppossums hang on to nipples in the mother's abdominal pouch

9

10

by hearing. It preys on small vertebrates and birds' eggs, and eats berries, carcasses and garbage. The opossum is quite successful in competition with placental mammals. It is fairly abundant in places, spreading constantly towards the north. It is a cautious animal, easily adaptable to changing conditions. It has an interesting habit of playing dead when approached by man, who hunts it for its valuable fur.

Native Cat (10) lives in south-eastern Australia and in Tasmania. Evolutionally, it represents the archetypal Australian marsupial. Its smooth fur is mostly yellow-brown or dark brown with white dots, and the thickly furred, non-prehensile tail has a white tip. The Native Cat frequents forests and open land, even in the vicinity of man. It is active at night, feeding on small vertebrates, insects and plants; it often catches sea animals on the shore. The female gives birth to six to eight young, but often some of the offspring die, as she has only six nipples.

Dasyurus quoll
Dasyuridae

Fat-tailed Dunnart (11) is one of the smallest Australian marsupials, measuring 76—120 mm and weighing only 10—15 grams. The unpadded, partly fur-covered feet, are well-adapted for jumping. The remarkably thick tail stores fat. The dunnart is a rare and protected Australian marsupial; it is widespread in the west where it lives in light forests, grassy savannas and rocky areas. It digs burrows or builds nests of grass and leaves in rocky crevices, under fallen trees or in thick bushes. It is active at night and moves on all four feet or jumps with the hind legs. The diet consists mainly of large insects, but it may include small mammals, for example House Mice. The young spend the first stage of their lives in the mother's

Sminthopsis crassicaudata
Dasyuridae

11

marsupium. Later the mother carries them on her body, in the manner of the opossum. The ecology of reproduction largely corresponds to that of small rodents. A series of litters is produced in the season of vegetational growth, the number being regulated by the weather and the abundance of food. In favourable years, ten young can be born in one litter. At low temperatures, or when the food is scarce, the species become torpid and thus save energy.

Marsupial Mouse (12)
Antechinus bellus
Dasyuridae

is a representative of the numerous order of evolutionally primitive and little specialized marsupials. It inhabits forest habitats of northern Australia and feeds on insects and small vertebrates. Most species lack a pouch; it may be replaced by a shallow fold of skin. The litter averages three to eight young, suckled for 3 months.

Tasmanian Devil (13, 14)
Sarcophilus harrisii
Dasyuridae

is a most curious predatory marsupial. However, there is nothing 'devilish' in the creature's looks or behaviour. The head is disproportionately large, the back reduced, the legs are short, stout and slightly bowed, and the tail quite long and bristly. The skin on the abdomen and head is visible and turns crimson when the animal is irritated. The strong-jawed mouth with 42 teeth, including long, dominant canines, can be opened wide to nearly a right angle. The Tasmanian Devil is a prototype of a predatory marsupial, although it has to feed on carcasses and slow-moving vertebrates because of its own sluggish and clumsy movement. It pushes the food into its mouth with the front feet, which is never done by beasts of prey. The devil is a nocturnal animal; it spends

The Tasmanian Devil can manipulate food with its front paws

the day hidden in empty rock clefts, in a nest of leaves and grass. In May, the female gives birth to two to four young, which for about 15 weeks remain attached to her nipples in a shallow marsupium opening to the back. They are suckled up to the fifth month, staying in the nest during that time. The devil used to be widespread throughout Australia: its elimination has been caused by dingoes and the aborigines. Nowadays, it is confined to Tasmania, where it has again become common in places, thanks to its strict protection.

Tasmanian Wolf (15), *Thylacinus cynocephalus* Dasyuridae

110—130 cm long, used to be the largest Australian predatory marsupial. Externally, it was remarkably similar to dogs or wolves, differing only in the thick, 50-cm-long 'kangaroo' tail, and the characteristic 13 to 19 transverse black-brown stripes, extending from the shoulder-blades to the root of the tail. This pattern gave it its English name Native Tiger. The considerable similarity to true beasts of prey is confirmed by the structure of the limbs (they tiptoe when walking) and of the skull, which reveals its marsupial nature merely by the number of teeth (46) and by the wide-opening jaws. Fossil findings bear out that over the past 10,000 years,

12

13

14

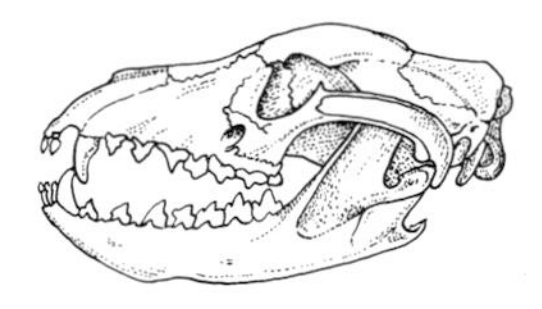

Skull of the Tasmanian Wolf resembles a canine skull

the Tasmanian Wolf had been distributed throughout Australia and New Guinea. White settlers, however, found it already restricted to Tasmania; its constant attacks on flocks of sheep had made it a widely pursued animal, which led to its complete extermination. Between 1888 and 1909, 2,184 rewards were paid for captured Tasmanian Wolves; the number of animals killed certainly surpassed that of rewards. This high rate later decreased and the last specimen living in the wild was killed probably in 1930, and those kept in zoological gardens perished soon afterwards. Although there were some unconfirmed reports of observation or traces of Tasmanian Wolves in the fifties and sixties, a special expedition undertaken in 1955 with the aim of verifying them, was unsuccessful. Except old zoological reports, all that remains are stuffed specimens or skeletons in various museums, a few photographs, and one film; on the basis of this material, it is supposed that the Tasmanian Wolf hunted at dusk and at night, capturing small kangaroos, birds, lizards, and exceptionally echidnas. It was shy and apprehensive, never attacked men and defied dogs only when in trouble. Females produced three or four young. They were kept for some time in the pouch, which opened to the back.

15

16

Numbat or **Marsupial Anteater** (16) is another rare and protected Australian marsupial. It is conspicuously coloured: six to twelve whitish stripes stretch transversely over the grey-brown back and a dark band runs along each side of the head. Its diet consists of ants and termites, dug out and collected with the long, extensible tongue. Dietary specialization probably influenced the form of the teeth: the jaws are lined with 50 to 56 tiny teeth. The Numbat lives in south-eastern Australian forests, and, unlike most marsupials, it is diurnal.

Myrmecobius fasciatus
Dasyuridae

17

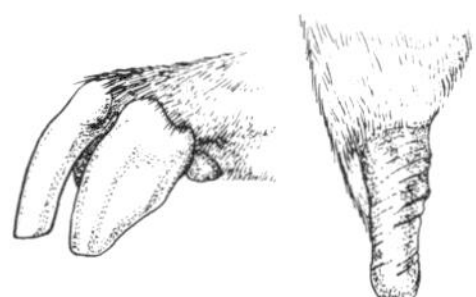
Front foot and tail of the Greater Marsupial Mole

18

Marsupial Mole (17) is evolutionally one of the most peculiar marsupials, differing enormously
Notoryctes typhlops from the basic type. In shape and size, it is almost identical with a real
Notoryctidae mole. However, the anatomy of its limbs is different, and it has a horny plate on the muzzle and horny rings on the tail. The short, thick fur has various shades of colour, from whitish to golden red. The Marsupial Mole lacks ear lobes, and its tiny eyes are often covered by a fold of skin. The female's marsupium opens to the back: usually only one young is born. The mole spends most of its life underground, coming to the surface only in rainy or flood seasons. It inhabits sandy and bushy lowlands of western and central Australia.

19

Koala (18, 19) is a relatively large marsupial, reaching 60 to 80 cm in length and, when adult, 16 kg
Phascolarctos cinereus
Phascolomidae

in weight. It feeds exclusively on leaves of eucalyptus trees, being thus one of the greatest food specialists among mammals. Ecologically, it represents the sloths in Australia. Its limbs are well-adapted for climbing trees, having long, sharp, slightly bent claws. A firm grasp is facilitated by the thumb and forefinger of the front feet being opposed to the other three fingers. Koalas are either solitary, or live in small groups. The mother carries her single cub in her pouch for 6 months, and then on her back for another 6 months. The deep marsupium opens to the back, which allows the infant at the end of lactation to feed on a mash of eucalyptus leaves, exuded by the mother from her rectum at certain hours of the day. This provides a gradual transition from milk to the unusual and not easily digested green food. A hundred years ago, koalas were extremely abundant in Australia, inhabiting all the available eucalyptus woods. They have become almost wiped out by senseless hunts for pleasure or for the valuable fur, by diseases and by tree-felling. The determined protection of the species in the past decades, combined with their re-introduction to places where they used to live, is helping to boost their numbers.

Hairy-nosed Wombat
Lasiorhinus latifrons
Phascolomidae

Skull of the Wombat exhibits certain rodent features

(20) remotely resembles a dwarfish bear. Its weight is 25—27 kg. It lives on plant food: leaves, berries, roots and tubers. This food specialization is associated with the formation of stout, digging feet, and particularly with extensive changes in the shape and functions of the skull, which is similar to the skull of a large rodent. The jaws bear two sharp incisor teeth with continually-growing roots; the five large cheek-teeth, with an almost flat gnawing surface, are divided from the incisors by a large gap called the diastema. Wombats are most active at night; during the day they hide in deep, self-dug underground burrows. They are solitary, except during the mating season. Females give birth to one, rarely two young, kept in the marsupium for several months. The Hairy-nosed Wombat is a resident of south-eastern and southern Australia.

20

21

Brush-tailed Phalanger (21) is the most abundant species of the phalangers. The size of a cat,
Trichosurus vulpecula
Phalangeridae

it has a long tail, covered in fur except for a prehensile spot on the underside of the tip. It lives in Australia except the northernmost area, and in Tasmania, occurring mostly in open country, but nowadays even around human settlements, in city parks and gardens. It is active at night, climbing trees, running on rooftops, and visiting gardens and garbage heaps. It feeds on leaves, flowers, shoots, fruits, garbage and insects. It is often hunted for its fur. It has a year-round breeding season. Usually only one young is born, and this is kept in the pouch for about 100 days. The life-span is approximately 13 years.

Spotted Cuscus (22) is an arboreal marsupial, about the size of a cat, inhabiting the humid
Phalanger maculatus
Phalangeridae

forests of New Guinea and the northernmost tip of Australia. It has a thick, curly coat, a round head with large, yellow eyes, a furry, prehensile tail with a naked tip, and sharp, hooked claws. There are several colour variants; the spotted one is the most common. The Cuscus spends most

22

of its life in trees; during the day, it sleeps in tree hollows or clutching to boughs. It is active at night, but moves slowly and lazily. The diet consists of fruits and leaves, but it also includes insects or birds' eggs and fledglings. The female's marsupium is well-developed; a single young is reared at a time.

Striped Phalanger (23)
Dactylopsila trivirgata
Phalangeridae

resembles a squirrel both in size and appearance, and is regarded to be the most beautifully coloured member of its family. It inhabits the rain forests of northern Australia and New Guinea. It feeds almost exclusively

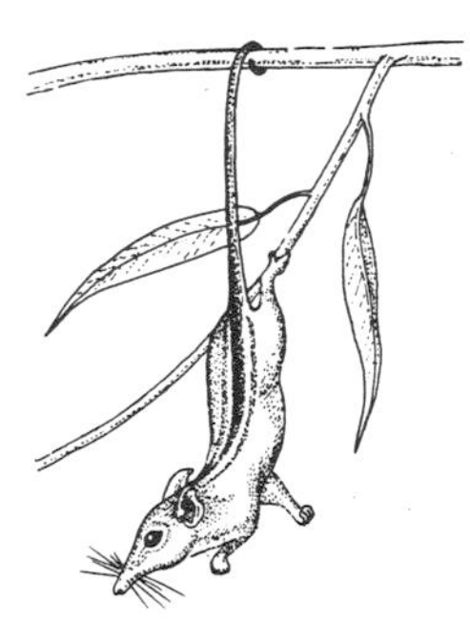
The Honey Phalanger (*Tarsipes spenserae*), a small relative of koalas, is a most interesting marsupial. It feeds on nectar and pollen

23

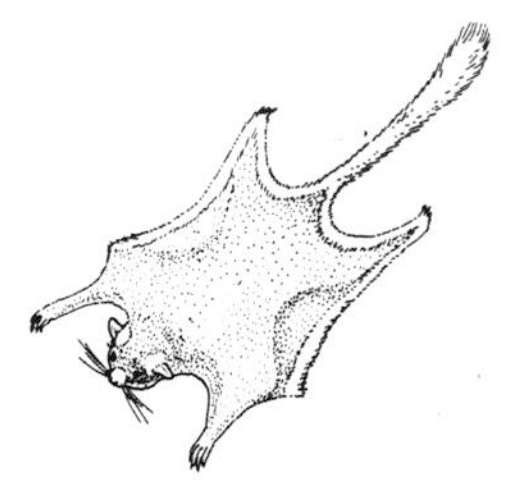

The Sugar Glider in a flying jump

24

on larvae of wood-boring insects. It searches for them in the manner of a woodpecker, drumming with its forefeet on the boughs and detecting the prey's presence by listening. Insects are drawn out either by the sharp incisors or by the conspicuously elongated fourth finger of the foreleg.

Sugar Glider (24) is one of the several arboreal marsupials which are capable of gliding flight,
Petaurus breviceps
Phalangeridae

thanks to a special fold of skin connecting the front and hind limbs and

25

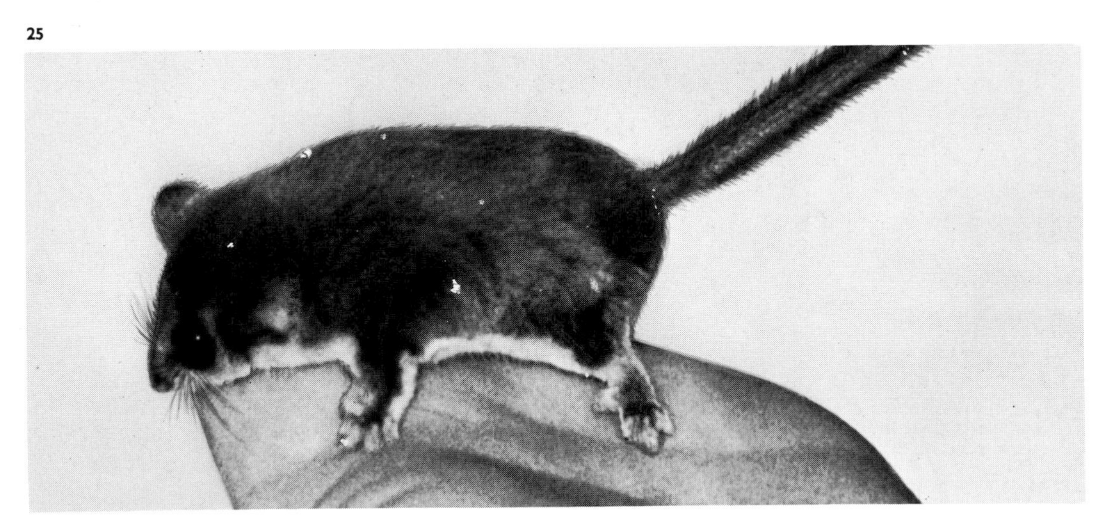

forming a simple flying membrane when the limbs are stretched out. The fold is hardly noticeable in climbing. The Sugar Glider can make a gliding jump up to 55 metres with the tail serving as a rudder. It feeds on insects and is very fond of sweet things (hence the English name 'Sugar Glider'). It is widespread in the forest areas of New Guinea, Australia and Tasmania, often occurring near human settlements.

Pygmy Glider (25), the size of a mouse, is a flying marsupial with large, black eyes, and a long, stick-like tail. It lives predominantly in eucalyptus forests of eastern Australia.
Acrobates pygmaeus
Phalangeridae

26

27

Female kangaroo giving birth

Kangaroos are a peculiar and well known group of Australian herbivorous marsupials. Adaptation to vegetarian food brought about many morphological modifications, particularly to the skull: for instance, the number of teeth, especially incisors, is decreased. Other characteristic features of kangaroos are the shortening of the forelimbs and the disproportionate development of the hind limbs, with domination by the elongated fourth finger, while the others have shortened or even disappeared, the second and third even fusing into one small finger. The long, muscular tail, thickened at the base, is used as an important organ of locomotion: it serves as a prop when the animal is standing, and maintains balance when it is leaping. The head is relatively small, with a protracted muzzle and long, 'hare-like' ears. Kangaroos are restricted to the Australian region: about 50 species exist today. They vary in size from rabbit-like creatures to animals as much as 2 metres tall and weighing over 90 kg. They are adapted for swift motion in long jumps, which is why the majority of them live in open flatlands. Like other herbivorous animals, they have a composite stomach for digesting plant food, assisted by symbiotic bacteria. The pre-digested food is chewed a second time.

Rufous Rat Kangaroo
Aepyprymnus rufescens
Macropodidae

(26) is not a typical kangaroo, with its 'rat-like' head and rather short ears, and with a not very great difference in the lengths of front and hind limbs. It belongs to the group of medium-sized 'rat' kangaroos and inhabits the steppes, forest-steppes and light woods of south-east Australia. When building a nest, it transports the material on its belly, supporting it there with its tail.

Rock Wallaby (27) is the best-known representative of the rock wallabies, medium-sized species
Petrogale penicillata
Macropodidae

with typical kangaroo features. They live in colonies in rocky areas, hiding under large boulders during the day, and going out late in the afternoon and in the evening. The places they frequent are recognizable by smooth paths on the stones. The Rock Wallaby is widespread all over Australia, from lowlands to mountains and from arid areas to the tropical zone.

Ring-tailed Rock Wallaby (28) is the most colourful of all rock wallabies. The body is ash-grey,
Petrogale xanthopus
Macropodidae

ears and limbs yellowish, a narrow, dark strip passes down the middle of its back, and its face and flanks show conspicuous whitish stripes. The long furry tail is adorned with dark bands. It is a rare inhabitant of southern Australia.

Red Kangaroo (29) is considered to be the largest living kangaroo species; males are up to
Megaleia rufa
Macropodidae

2 metres tall and weigh as much as over 90 kg. This is one of the five species of the so-called 'large kangaroos', best adapted to the life in grassy Australian savannas, and representing the final stage in the evolution of herbivorous Australian marsupials. Ecologically, they represent grazing ungulates. Being good runners, they can reach a speed of 80 km/hour

28

Jumping kangaroo

29

over a short distance, and can jump as far as 9 metres. However, they cannot run for long and soon tire. Originally, they were very abundant in Australia; later, they became competitors for food with sheep, were pursued and decreased in number. They are the kangaroos most usually represented in zoological gardens.

Black-faced Kangaroo or **Western Grey Kangaroo** (30) is another large kangaroo, closely related to the Red Kangaroo. It is resident in forest areas of the coastal belt of south-eastern Australia.

Macropus fuliginosus
Macropodidae

30

Swamp Wallaby (31) is a typical representative of a numerous group of so-called 'large wallabies', differing from the kangaroo genus *Macropus* by their smaller size. With some exceptions, wallabies also inhabit open plains, but they settle only in places where a safe shelter in shrubberies or under trees can be found. They visit open grassy spaces only to graze, usually in late afternoon or

Wallabia agilis
Macropodidae

31

32

in the morning. Together with large kangaroos, they provide the typical kangaroo community of the Australian region, and occur quite frequently in places. Since they are regarded as competitors with sheep and cattle, they are hunted excessively. They provide canned meat for dog food, and the fur of certain species is fashionable.

Parry's Wallaby (32) lives in herds in eucalyptus woods of south-eastern Australian hills. Its skin serves as material for manufacturing toy koalas, sold as souvenirs.
Wallabia parryi
Macropodidae

Tree Kangaroo (33) is one of the six peculiar species of kangaroo which during the course of evolution abandoned life on the ground, and totally adapted themselves to an arboreal life. They also have a substantially modified appearance. Both pairs of limbs are of approximately the same size. Their hind feet move independently (walking), while true kangaroos move them simultaneously (leaping). The long tail is less thickened at the base. They are medium-sized, about 70 cm long, weighing up to 11 kg. Their haunts are the rainforests of New Guinea and forests of the northernmost tip of Australia.
Dendrolagus matschiei
Macropodidae

33

New Guinea Mountain Wallaby (34) lives in the mountainous virgin forests of south-eastern New Guinea. It closely resembles tree kangaroos, but moves on the ground. The muzzle is flat and bare, the forelimbs are strong and the fur of the neck turns forwards.

Dorcopsis macleayi
Macropodidae

34

Chapter 3 INSECT-EATING MAMMALS

Insect-eaters (Insectivora) are generally considered to be the oldest order of placental mammals. Many of their features illustrate that, during their long evolution, they have differentiated little from the primitive types of Cretaceous mammals known from fossil remains. They remained small, with a great number of primitive, tiny, sharp-pointed but little differentiated teeth. The limb structure is primitive in most forms, without any specialization. Almost all insect-eaters have five-fingered feet, and they are plantigrade (walk on the entire sole). The small and flat cranium, which often lacks the zygomatic arches, contains a relatively small and simple unconvoluted brain. A pointed, often mobile muzzle is typical of most species. Otherwise, insectivores are a heterogenous group which can not easily be characterized by special common features. This is evidence that, since prehistoric times, they have developed along several parallel lines which later became sufficiently differentiated to merit classification into individual orders in the opinion of some zoologists. This is true, for example, of the unique elephant shrews and to some extent of the golden moles. The peculiar group of tree shrews is sometimes classified with the insectivores, sometimes with primates.

Despite their ancient origin, contemporary insectivores constitute a rather successful group, for they have colonized almost all regions of the world (with the exception of polar areas, Australia and a large portion of South America). There are about 400 forms and after rodents and bats, they rank as the third most numerous order of contemporary mammals.

Cuban Solenodon (36)
Solenodon cubanus
Solenodontidae

is one of the two living members of an ancient and previously widespread family of primitive insectivores of North America. It is the largest representative of the order, measuring 50–60 cm, of which 20–25 cm is the

36

37

tail length, and weighing more than 1 kg. It resembles a strange 'cross' between a rat and a large shrew. The muzzle extends into an elongated snout, the 'rat-like' tail is long, and the long feet have digging claws. The body is covered by a thin, bristly coat, brown-black on the back and pale on the head. It was thought to be an extinct species, but several years ago, it was captured unexpectedly in eastern Cuba and the Havana Zoo has been keeping it as a rarity. A related species, *S. paradoxus*, lives in Haiti and is not so rare.

Otter Shrew (37)
Potamogale velox
Potamogalidae

is one of the largest insectivores. Its body measures 29–35 cm, and its tail almost equals the body length. It resembles an otter to such an extent that nobody would take it for an insectivore. It is completely adapted to life in water: it swims and dives perfectly. The long, dorso-ventrally flattened tail is the main organ of locomotion: its feet are not webbed. It feeds on crustaceans, reptiles and fish, caught in the water, usually at sunset. By day, it hides in its burrow which has an underwater entrance. The litter usually consists of two young. The Otter Shrew lives near forest brooks and still waters in western and central equatorial Africa; two of its smaller relatives were recently found to inhabit similar regions.

Tenrec (38) with its spiny coat resembles a hedgehog. It is found all over Madagascar, mainly in dry surroundings. As in the majority of tenrecs, its body temperature is not very constant and it hibernates in winter. It belongs to an independent family, Tenrecidae, regarded as one of the most primitive groups of living placental mammals. They have survived only in one place in the world, Madagascar. This isolated occurrence testifies to their ancient origin. In many features, tenrecs have hardly progressed beyond the evolutionary level of primitive marsupials, and they are little differentiated from original types of Cretaceous placental mammals. The isolation of Madagascar enabled them to diversify into many forms (there are 31 known species), and settle in various habitats. They include species resembling shrews or hedgehogs, and even species with webbed feet. The Tailless Tenrec (*Tenrec ecaudatus*), the largest and best known species, represents an early fur-covered type: its body length is 26–35 cm. The female usually has about 12 to 16 young at a time, but one litter was found to comprise 31 offspring. The developing young are furnished with erect quills on their backs: subcutaneous muscles make the quills vibrate and emit sounds, serving probably as a means of communication.

Echinops telfairi
Tenrecidae

The Tailless Tenrec ***(Tenrec ecaudatus)*** **is the best-known member of the tenrec family**

38

39

Common Mole (39) is excellently adapted for underground life. It has a cylindrical body with short, thick fur and strong forefeet in the shape of backward-turned scoops. The outer ears are protected by a narrow fold of skin, and the eyes are tiny, sometimes overgrown by soft skin. Most of its life is spent in underground passages which are constantly being carefully cleaned and rebuilt. The mole's diet consists of insects, their larvae, worms, and

Talpa europaea
Talpidae

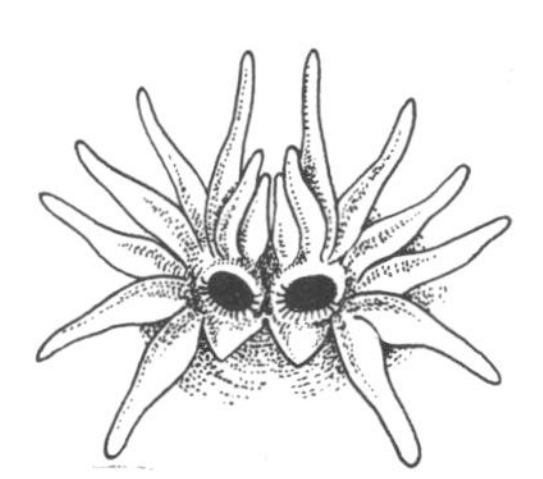

Fleshy outgrowths on the muzzle of the American Star-nosed Mole ***(Condylura cristata)***

40

41

small vertebrates, mostly frogs. It can even store up 'live preserved food', that is paralysed earthworms, bitten through the nerve centre. The extremely greedy moles destroy many pests, but they are pests themselves — capturing earthworms, damaging plant roots and throwing up molehills. They are widespread all over Europe, except the northernmost areas, and in the east range as far as the Urals. The mole family includes about 20 underground and semi-aquatic forms, distributed in the Eurasian temperate zone and in North America.

Russian Desman (40)
Desmana moschata
Talpidae

is a close relative of moles, although it looks substantially different. It is a semi-aquatic animal, somewhat larger that the Common Mole, with a thick, close and waterproof coat. Its body measures 18—21 cm, the tail is 17—20 cm long and the weight is 250—400 grams. The toes of the hind feet are webbed, and the laterally flattened tail, with a sparse covering of hair, is used as a rudder. The muzzle is elongated into a tiny, naked, prehensile trunk. The musk gland located at the base of the tail exudes a foul-smelling secretion. The desman lives on banks of slow-flowing brooks and water reservoirs with thick surrounding vegetation. The diet of molluscs, worms, insects, fish and frogs is sometimes supplemented by plants. The burrows dug in banks have underwater entrances;

42

the grass-lined nest is located above water-level. The Russian Desman used to be hunted for fur and was almost exterminated. At present, it is becoming more abundant thanks to vigorous protection and re-introduction to some places. It is confined to a relict territory in south-eastern Europe, in the basin of the Don, Volga and Ural rivers. Its smaller relative, the Pyrenean Desman (*Galemys pyrenaicus*), occurs in the Pyrenees.

Grant's Desert Golden Mole
Eremitalpa granti
Chrysochloridae

(41) is a member of an independent family of golden moles, an ancient group of 15 species of insectivores, inhabiting southern and eastern Africa. Their mole-like appearance is related to the underground way of life and does not reflect any closer evolutionary relationship. They differ from moles in the anatomy of the forelimbs, which are four-toed, narrow, and supported by an ossified tendon. They do not dig with the whole front paw like moles, but only with the strong claws of two middle toes. Grant's Desert Golden Mole inhabits the deserts of southern Africa. It is the smallest of all golden moles, weighing a mere 15 grams. The individual species of golden mole differ in size and in colouring of the shiny, metallic coat.

North African Elephant Shrew (42) and the other 30 species of elephant shrews resemble jerboas in both appearance and way of life. The family represents a related line of insectivores, which has developed separately and created remarkably specialized forms. Their most conspicuous features are the long, thin and naked limbs adapted for jumping, the elongated, pointed muzzle and the long, mouse-like tail. They inhabit dry, grassy and rocky areas throughout Africa and Zanzibar. Unlike most insectivores, they are active during the day. They forage for invertebrates, mainly insects, and feed partly on plants. Certain species have a specialized diet of ants and termites, others even attack small vertebrates. The North African Elephant Shrew, the best known in the family, lives in semi-deserts and deserts of northern Africa.

Elephantulus rozeti
Macroscelididae

43

44

European Hedgehog (43) is the most popular of all the insectivores; it is a member of the hedgehog family, which includes some 15 species inhabiting Eurasia and Africa. Its most outstanding feature is the coat of several thousands of tough, prickly spines, which are in fact modified hairs. These are an ingenious defensive device, together with the hedgehog's habit of curling into a ball when danger threatens. Hedgehogs go out at dusk and their foraging trips are disclosed by rustling, trampling and snuffling sounds. They consume large quantities of insects, worms, molluscs, tiny vertebrates and plants, and their diet also comprises venomous snakes, daringly

Erinaceus europaeus
Erinaceidae

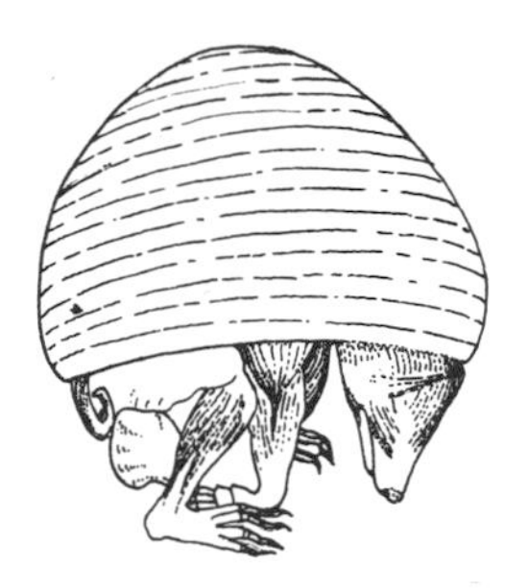

Strong skin muscles enable a hedgehog to roll into a ball (seen without spines)

and dexterously overpowered. Hedgehogs are not, however, quite immune from snake venom as is generally thought. Birds' eggs and fledglings are a delicacy for them, which makes them unwelcome visitors to pheasantries. Still, they are useful and protected in many countries. Many hedgehogs are unfortunately killed by cars and by chemicals used in pest control. Hedgehogs, mainly males, live solitarily in relatively small territories. They are often seen at forest borders, in gardens and parks. They prosper in the wide range of environments associated with man. In winter, from November to March, they hibernate in shelters lined with dry grass and leaves. The female bears two to ten offspring once or twice a year. Their life expectancy is 8–10 years. A closely related species, *Erinaceus roumanicus*, differing in colour, also lives in Europe.

African Hedgehog (44)
Hemiechinus auritus
Erinaceidae

is smaller than its European counterpart, more lightly coloured, and has larger ear lobes. It is a resident of steppes and semi-deserts of eastern Europe, western and central Asia, India and Egypt.

The shrew family has an almost worldwide distribution with the exception of polar regions, Australia, and a part of South America. It embodies about 300 species, that is three-quarters of all recorded insectivores. All of them are small: they include the smallest living mammals, for example Savi's Pygmy Shrew (*Suncus etruscus*) which weighs a mere 2 grams. Shrews are agile animals with a pointed muzzle, thick velvety fur, beadlike eyes and short ear lobes, which in some species are

45

46

entirely covered by fur. They have a tiny, flat skull without zygomatic arches, and 26 to 32 small, little differentiated teeth. Their small feet have retained the original five-toed form without any specialization. Many predators avoid them because of the scent glands situated at the sides and base of the tail. Their diminutive size and consequently a relatively large body surface area cause high thermal losses which are made up by huge food consumption, daily surpassing their own body weight by one-half. Without food — for instance in a trap — they perish very fast. In normal circumstances, when sheltered in their underground nests, they are very resistant and flourish even in mountains or northern areas. Together with small rodents, some shrew species are the most abundant mammals.

Pygmy Shrew (45) is a typical representative of the genus *Sorex*. All its members have the tips of their 32 tiny teeth coloured dark red. They breed from spring to autumn and have one or two litters numbering five to eight young. Like all shrews, the Pygmy Shrew is short-lived, dying after 16 to 18 months in the wild. It is one of the smallest mammals, measuring only 45 to 60 mm and weighing 3 to 5 grams. It inhabits various habitats, mostly thickets, both in lowlands and mountains. It is widespread over Europe, except the western Mediterranean and Iceland, and in central and southern Siberia, Japan and China. The related Alpine Shrew (*Sorex alpinus*) (35) lives as a relict in some European mountain regions.

Sorex minutus
Soricidae

Eurasian Water Shrew (46) is one of the two representatives of the genus *Neomys*, confined to Europe, Asia Minor, and Siberia as far as northern China. It is a large shrew, weighing 10–20 grams. It keeps close to rivers, stagnant water and damp places. Although not much adapted for aquatic life, it is a good swimmer and diver and can run on the river bed. The fur is waterproof and the hind feet are rimmed by elongated tough bristles, replacing webs. Another row of these forms a keel on the underside of the relatively long tail. It is active day and night, searching for food mostly in water, but the prey — insects and larvae, small crustaceans, molluscs and fish fry — are eaten on land. Water shrews reproduce from spring to autumn, and have usually two to four litters yearly with three to eleven young. The closely related *Neomys anomalus*, more abundant in the southern parts of the area of distribution, is less dependent on nearby water.

Neomys fodiens
Soricidae

Lesser White-toothed Shrew (47) represents the most diverse genus of shrews, including some 150 species. As one of the few members of the genus, it penetrated in Europe as far north as northern France and central Germany. It prefers warmer sites in open, cultivated landscapes: forest borders, slopes with shrubberies, fields and the surroundings of human settlements. Sometimes it follows brooks and lines of communication up to forests and mountains. In winter, it regularly moves to sheltered places: haystacks, piles of stones and even houses, where it gets caught more often than other shrews in traps set for house mice. White-toothed shrews differ from other shrews morphologically and in their way of life. Their main characteristics are: light grey colouring, white teeth, a triangular head with prominent ear lobes, a rather short tail covered by a short coat interspersed with individual long hairs. In recent years some species of European white-toothed shrews have been successfully bred in captivity. Their biology is therefore quite well-known: following a gestation lasting 28 days, the females produce two to six blind and naked offspring. Generally, there are two to four litters annually. White-toothed shrews are mostly found in Africa and central and southern Asia. In Europe, they are most abundant in the Mediterranean.

Crocidura suaveolens
Soricidae

47

48

The photograph shows **an unorthodox way of leading the young,** developed probably in most white-toothed shrews. The family follows the mother in a single file, each youngster clutching the preceding one by the hair at the root of the tail. Thus, in case of danger, the female can take her offspring, blind and incapable of orientation, to shelter.

Bicolor White-toothed Shrew (48) is recognized by its lightly coloured belly which contrasts sharply with the grey sides (unlike the Lesser White-toothed Shrew, where the colouring of the sides grades into the light colour of the abdomen). In Europe, it inhabits the same sites as the Lesser White-toothed Shrew but is less abundant and its distribution is rather patchy. Biologically, it differs little from the previous species, but is probably more warmth-loving.

Crocidura leucodon
Soricidae

The subfamily Crocidurinae includes the genus *Suncus,* comprising besides the smallest living mammal, Savi's Pygmy Shrew, found in the Mediterranean, western and central Asia and India, also the large House Shrew (*Suncus murinus*), adapted to commensal life in human dwellings.

The House Shrew ranges from eastern Africa over Arabia and southern Asia to the Philippines. It is one of the food-storing animals. Its close relative, the African Shrew *Praesorex goliath*, the largest known shrew, measures 15—18 cm, and the tail is about 11 cm long.

Piebald Shrew (49) *Diplomesodon pulchellum* Soricidae

has unusual colouration for a shrew: a long, oval spot is located in the middle of the greyish-black back, and the underparts, feet and tail are white. It lives in a limited territory of the central Asian and Kazachstan deserts, being in fact the only insectivore species totally adapted to the conditions of a sand desert. Its way of life is little known: it is mostly active at dusk and at night. During the day it hides underground, either in self-dug burrows or others abandoned by various rodents. In addition to insects, it feeds on small desert lizards. The female has approximately three litters a year. The Piebald Shrew is small, about 5—8 cm long with a short tail and paws fringed with tough bristles, facilitating movement in sand.

49

50

Chapter 4 FLYING MAMMALS

The only mammals endowed with the power of active flight are bats. Their wings are nonetheless built on an entirely different principle from those of birds: they are transparent membranes — double layers of almost hairless skin — stretched between the elongated front limb bones, the hind limbs and the tail. As regards other external features as well as internal anatomy, bats do not differ substantially from insectivores. They are also closely related to the primary types of placental mammals. We know nothing about their evolution or about the development of their capacity for flight, but the oldest fossilized remains, discovered in Eocene sediments, strongly resemble the contemporary species.

Bats have yet another peculiarity, which has since long ago intrigued man's imagination and was satisfactorily explained only recently. They are capable of safe flying and even insect-hunting in pitch darkness: their special method of orientation in space, based on echolocation, is a sort of sonar device. This is present in all typical bats, and even in some fruit bats. Their sensitive hearing detects echoes of high-frequency sounds, emitted by mouth or nose, and this perception enables them to avoid obstacles or locate flying prey.

Thanks to their capacity for flight, their nocturnal life-style and other ecological and physiological adaptations, bats have become a very successful group, ranking second after rodents in number of species (about 900 are known). Except in polar areas and remote oceanic islands, they can be found throughout the world. They are divided into two extensive suborders: fruit bats (Megachiroptera) and typical bats (Microchiroptera).

The suborder of fruit bats has only one family (Pteropidae) with some 150 species distributed in subtropical and tropical zones of Africa, Asia and Australia. They are distinguished from typical

51

bats by large eyes, ear lobes without a tragus, a claw on the second finger and by the absence of a tail. Most species are guided by their sight and smell, only the genus *Rousettus* has developed a simple sonar. They spend the day sleeping in large colonies in tree tops, or even under roofs or in caves. Most species, especially the larger ones, feed on soft fruits; the smaller ones also on nectar and pollen, thus serving also as pollinators of certain plants. Some species of the genus *Pteropus* have a wing-span of up to 1.7 metres and weigh about 1 kg; contrary to this, some other species weigh a mere 15 grams.

Short-nosed Fruit Bat (50) is one of the smaller fruit bats of the Indo-Malaysian region. It lives in small colonies of two to twelve individuals. In the daytime it hides in caves, mine shafts, under the eaves of houses and in palm branches. Sometimes it even makes an ingenious shelter in the middle of a cluster of fruit, where it bites out a niche. Like most fruit bats, it feeds on juicy fruits, sometimes on nectar, in search of which it often travels long distances.
Cynopterus brachyotis
Pteropidae

Wahlberg's Epauletted Fruit Bat (51) occurs in western and eastern Africa. It lives in small colonies, spending the day resting in trees.
Epomophorus wahlbergi
Pteropidae

Straw-coloured Bat (52) is widespread in Africa as far north as Egypt, in Madagascar and in the south of the Arabian peninsula. It often forms large colonies which may be found sleeping in tall trees, or even parks and city gardens. Early in the evening, the colony sets off to forage for food and often returns at dawn. The diet consists of the soft fruits of various trees.
Eidolon helvum
Pteropidae

52

53

54

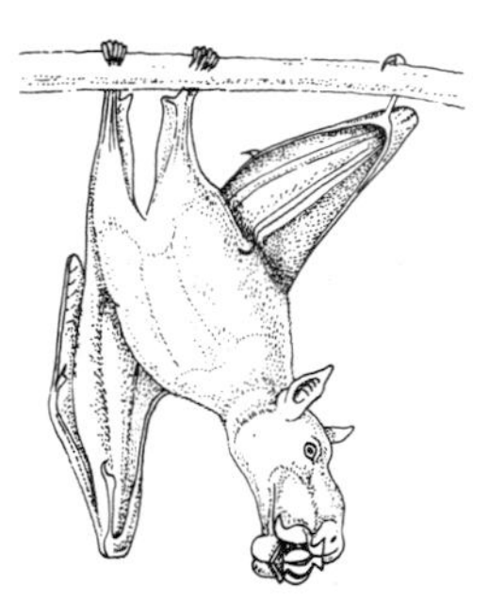

African Hammerhead Fruit Bat *(Hypsignathus monstrosus)* **is one of the most bizarre Fruit Bats**

Greater Horseshoe Bat (53, 54) is a true bat of the suborder Microchiroptera which includes 750 species divided into 16 families. The majority of them are small animals weighing 3—50 grams. They are mostly insectivorous — as is evident from their dentition, formed by a large number of sharp little teeth. Other species live on fruits, nectar, pollen, blood or small vertebrates. They are even better adapted for flying than fruit bats; unlike them, they lack the claw on the second toe, their ear lobes are more complicated and many families have membraneous outgrowths on the muzzle. Their orientation is based on echolocation; both sight and smell are poorly developed. They originated in the tropical zone, but are now found everywhere. Living often in colonies, they hide in tree hollows, underground caves and human dwellings. They usually have one or two young every year. They are noted for a long life-span: up to 25 years. Insectivorous species of temperate zones hibernate.

Rhinolophus ferrumequinum
Rhinolophidae

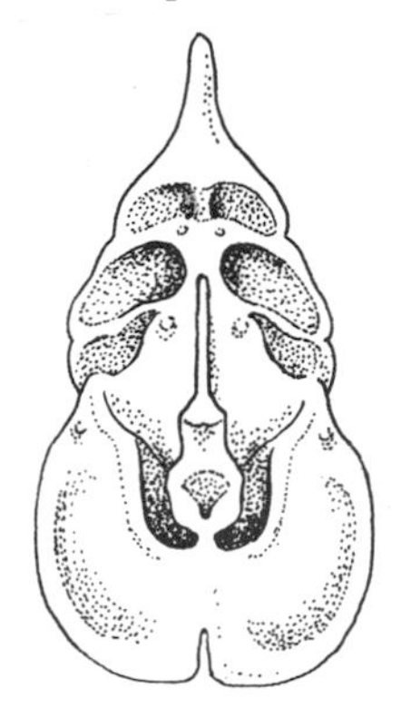

Membranous outgrowth on the muzzle of the Horseshoe Bat

Greater Horseshoe Bats can be recognized at once by the bizarre membraneous outgrowth on their muzzle which serves as a regulator of sounds emitted by the fine sonar (the sounds are emitted through the nose). They have a relatively short tail, large, membranous, pointed ears, bent sideways and without a tragus. Near the genital orifice of females are special skin nipples which the young bats grasp during transport. The Greater Horseshoe Bat is one of the largest members of the family (56—69 mm long and weighing 17—28 grams). It lives in colonies in

55

56

warm caves in the Mediterranean, in western and central Asia as far as Japan, and in the north as far as central Europe and the British Isles. When resting or hibernating, it hangs suspended from the ceiling, attached by its claws and wrapped in the soft flying membranes; it never touches its neighbour. Like most true bats, it has only one young at a time.

Lesser Horseshoe Bat
Rhinolophus hipposideros
Rhinolophidae

(55) is a miniature of the previous species; it weighs only 4 to 9 grams and is believed to be one of the smallest bats. It is a resident of warm parts of Eurasia and northern Africa, extending further north than all other horseshoe bats. In central Europe, its numbers have decreased

57

during the last years as a result of the increasing use of chemicals and of disturbances to its sources of shelter. It was originally a cave dweller but in central Europe it now also inhabits attics and descends underground only in winter, which it spends in colonies of several hundred individuals. In summer, it sets up smaller colonies.

Heart-nosed False Vampire
Megaderma cor
Megadermatidae

(56) lives in Africa, from Ethiopia to Tanzania. Together with the other four species of the family, it is called a 'false vampire' because of its resemblance to true vampires. It has membranous outgrowths of

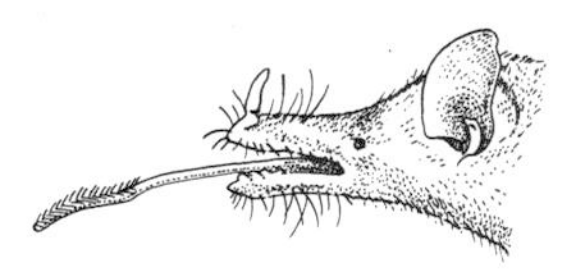

Some South American vampires feed on nectar and pollen. The drawing shows the head and protruded tongue of a vampire of the genus *Choeroniscus*

58

differing shape on the nose; the large ears have grown together in the middle for nearly half their length and the tragus is split. Megadermatidae, however, are quite unrelated to the true South American vampires of the vampire family (Desmodontidae). They live on insects; larger species on small vertebrates, and some are even cannibalistic. During the daytime, they sleep in caves, on buildings, in rocky clefts and in trees, usually in colonies of 50 to 100 members. The family is widespread in southern and eastern Africa, south-eastern Asia and Australia.

59

Vampire Bat (57) is a member of a smaller but notorious family of American leaf-nosed bats
Desmodus rotundus
Desmodontidae

with an extraordinary food specialization: they feed on the blood of warm-blooded vertebrates, licking it with their long, tubular tongue from wounds made by the knife-like incisors. They are well adapted to this

60

method of feeding by having a flat, bulldog-like snout and specialized teeth; furthermore, they can move agilely on all four feet, using the forearms of their folded wings as forelimbs. This movement, quite exceptional among bats, allows them to chase their host on the ground and crawl without trouble over its body. At daytime, vampire colonies rest in caves, hollow trees or in buildings, setting out to hunt at dusk. They attack mostly cattle and wild quadrupeds, and only rarely people or dogs

61

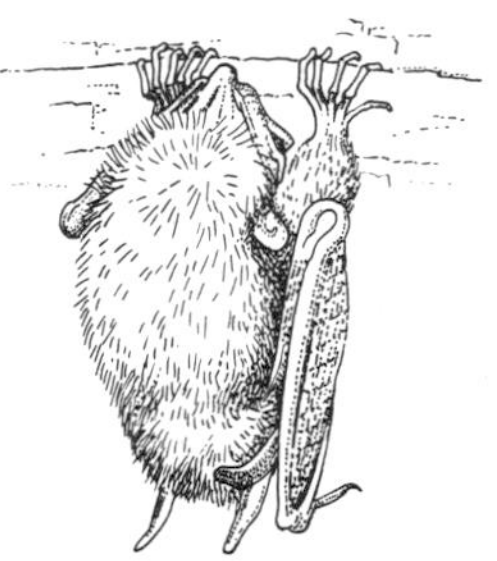
Hibernating Long-eared Bat has its long ears folded under its wings

62

(dogs can hear the ultrasounds, which wake them). Vampires cause great damage in South American cattle-breeding, not so much by weakening cattle when lapping blood, but by transferring the viruses of dangerous diseases, especially rabies. They are restricted to the tropical and subtropical zones of Central and South Americas, as far north as Mexico.

Long-eared Bat (58) is the best known species of bat in Europe, representing the huge family of mostly temperate-zone insectivorous bats in school textbooks. It is a small animal with conspicuously long, wrinkled and transparent ear lobes, which are folded under the wings when the bat is resting. It is widespread throughout Europe, temperate Asia and northern Africa, and it is common in every suitable habitat. In summer, the females with their young form smaller colonies in attics, hollow trees and birds' nest-boxes; winter is spent individually underground or in sheltered places above the ground. Hunting takes place after sunset; the bats fly slowly but skilfully around tree tops and catch mostly moths, although they can also pick up insects from leaves.

Plecotus auritus
Vespertilionidae

Common Noctule Bat (59) inhabits all of Eurasia except the northernmost parts and is quite abundant in some regions. In summer, it hides mostly in hollow trees in mixed and deciduous woods, in parks and avenues of trees. The females and their young form colonies of ten to 50 specimens, males are solitary. They set out before dusk to hunt, flying swiftly at high altitude. In winter they hibernate in hollow trees, rocky crevices and attics, often in large colonies. Northern bat populations fly south to hibernate, sometimes 1,000—2,000 km away from their summer sites.

Nyctalus noctula
Vespertilionidae

Mouse-eared Bat (60) was originally an inhabitant of warm caves in the Mediterranean; when it started to use shelters in attics of churches, castles etc., it spread throughout Europe up to northern France, central Germany and central Poland, and also to the Ukraine, the Caucasus, western Asia and Asia Minor, and northern Africa. In summer, females and juveniles form large colonies of several hundreds, but they hibernate in smaller groups or individually. The Mouse-eared Bat is the largest member of the genus; it has 66 species and ten of them live in Europe.
Myotis myotis
Vespertilionidae

Long-winged Bat (61, 62) has tiny ear lobes which do not protrude beyond the outline of the head. It is a warmth-loving cave-dweller. It is widespread in a vast area from south-western Europe and Africa to New Guinea and Australia. It lives in caves in large colonies of several thousand members, an example of which is shown in the accompanying photograph (62).
Miniopterus schreibersi
Vespertilionidae

Free-tailed Bat (63) is a little known south-African member of the peculiar family of Molossidae, represented by its 80 species in warm areas of all continents. They are easily recognized by their wide, wrinkled mouth, broad and thick ear lobes and mostly by the tail, which is excluded from the flying membranes (hence 'free-tailed bat'). Their long, narrow wings make them the best flyers among bats (they reach a speed of 60—95 km/hour). Some species, for example the Mexican Free-tailed Bat (*Tadarida brasiliensis*), assemble in large colonies of several millions of members in the south of the USA and in Mexico. Their droppings — guano — are used as an excellent fertilizer.
Sauromys petrophilus
Molossidae

63

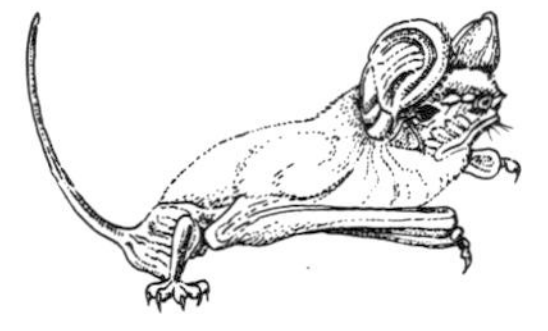

European Free-tailed Bat (*Tadarida teniotis*) is the only European member of the family of free-tailed bats (Molossidae)

64

Chapter 5 MAN'S CLOSEST RELATIVES

The order of primates (primates) constitutes a diverse group which, since Linnaeus' days, has included Man as a biological species, together with his extinct or surviving animal ancestors. Even the modern representatives reflect the long and complicated evolution of primates from types demonstrating a close relationship with the ancient insectivores (tree shrews), through ancestral prosimians, New World and Old World monkeys, to great apes and Man. To give a precise and complete characterization of the whole order is hardly feasible: in fact, its scientific name, supposed to stress the 'priority' and 'superiority' among mammal orders, does not quite correspond to reality. In some directions — in terms of brain development and mental abilities — primates have reached the undisputable peak of evolution within their class, but in terms of other physical features, for example anatomy of the limbs, dentition, etc., they have not developed much beyond the original placental animals, and they have been overtaken by some more 'modern' orders, for instance by various ungulates. As a result of this, primates are now envisaged as situated rather toward the beginning of the natural system of mammals than at its end, as was the case in the recent past.

The diversity of primates and their varying degree of evolutionary advancement is reflected in their division into the two suborders: prosimians (Prosimii) and monkeys, apes and Man (Anthropoidea). Superficially, these groups seem to have very little in common. On the other hand, a layman will fail to notice the basic differences between the two subdivisions of the suborder Anthropoidea — New World monkeys (Ceboidea) and Old World primates (Cercopithecoidea) — in spite of the fact that any similarities have resulted more from like ways of life than from close relationship. When imagining a typical primate, one usually has in mind a member of the last superfamily (Hominoidea), including, besides humans (Hominidae), also gibbons (Hylobatidae), and especially the great apes (Pongidae), that is Chimpanzees, Gorillas and Orang-utans.

Common Tree Shrew
Tupaia glis
Tupaiidae

(65) has become an object of interest to all zoologists in the last few decades. It was discovered that its original classification with the insectivores was not justified in view of many features pointing to its close relation to prosimians. Consequently, it is now classified either in the order of primates, or in an independent order of Scandentia, exemplifying possible lines of prehistoric evolution from insectivores to primates.

65

66

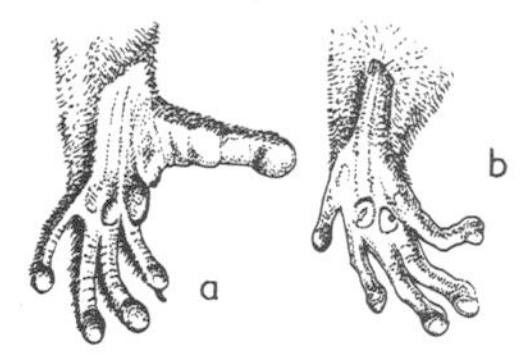

Shape of hand (a) and foot (b) in lemurs

The progressive characteristics of tree shrews are: large and developed brain, partly closed-off eye sockets, pads on the paws, and peculiarities of internal anatomy. Tree shrews live in forests of south-eastern Asia from India to Indonesia and the Philippines; they form about 18 known species. They are extremely agile and lively, being active mostly in daytime. Most of them are arboreal, that is they live in trees. Their diet consists of insects and invertebrates, partly of fruits. Litters average one or two, exceptionally four young.

Ring-tailed Lemur (66), one of the best known lemurs, belongs to the suborder of prosimians.
Lemur catta
Lemuridae

The family of lemurs is nowadays strictly confined to Madagascar and the Comoro Islands. They became the first evolutionary line of Lemuroidea, being probably direct descendants of those prosimians which lived in the Tertiary Era both in North America and Europe, and which

The Ring-tailed Lemur basking in the sun

were the ancestors of all higher primates. *Microcebus murinus*, a little larger than a mouse, is the smallest species and the Indri (*Indri indri*), measuring about 100 cm, is the biggest. All lemurs have thick and conspicuously coloured fur, a long, bushy tail and flat nails except on the second toe of the hind foot.

Incisors and canines of the lower jaw are flat, forming a comb-like ridge. The brain is simple, but relatively larger and better developed than in the preceding mammalian orders. Most lemurs are inhabitants of virgin forests and their increasing scarcity is being brought about by the disappearance of Madagascan forests. The Ring-tailed Lemur is distinguished by its long, bushy tail, striped black-and-white. It is a diurnal species and lives on the ground, but is an able tree-climber and jumper. Groups of 5 to 20 individuals live together on rocky dry places in southern Madagascar. They feed chiefly on wild bananas and figs, holding food with their hands. Females give birth to one, rarely two young and suckle them for a long time. All lemurs have black-coloured glands on the forelimbs. They use the excretions for constantly scenting their skin and marking their territories.

Mongoose Lemur (67)
Lemur mongoz
Lemuridae

is the only lemur species occurring on the Comoro Islands. It is arboreal, active at dusk and at night. Its biology does not differ substantially from that of other lemur species.

Ruffed Lemur (68)
Varecia variegata
Lemuridae

has thick, long, waterproof fur which forms a high collar around the neck. It inhabits the rain forests of northern Madagascar, where it is nocturnal and lives in small groups. It is the only lemur that builds fur-lined nests in trees, using leaves and twigs. The young are reared here.

67

68

Aye-Aye (69) has conspicuously long fingers with tiny nails only on the thumbs. The middle finger of the front foot is very slender and long; its function is to pull out woodworm larvae, and pulp from bamboo stalks and nuts, which are the Aye-Aye's staple foods. When searching for insects, aye-ayes knock on the wood, listen with their ears and extract the prey using the gnawing teeth and the long finger. Today, they live only in the scarce remains of the virgin forests of northern Madagascar. Some time ago, they were installed on a nearby island in a special preserve.

Daubentonia madagascarensis
Daubentoniidae

69

The Aye-Aye *(Daubentonia madagascarensis)*

70

Slender Loris (70) represents the second evolutionary line of prosimians, the so-called lorisoids (Lorisoidea). It lives in the virgin forests of India and Sri Lanka. Hands and feet are prehensile, the thumbs opposed to the other fingers, which are vestigial. It lives exclusively in trees, moving slowly and apparently hesitantly along the branches; but it is fast when faced with danger. It feeds mostly on insects, but also gathers fruits, leaves, birds' eggs and small vertebrates. It is a nocturnal animal. The territory is demarcated by urine, spread with the paws and transferred from tree to tree. Following a 5-month gestation, the female produces one or two young and carries them until they grow up. Adults are solitary or live in pairs.
Loris tardigradus
Lorisidae

Slow Loris (71) is a close relative of the Slender Loris. It lives in southern Asia, from India to the Philippines. It is even more sluggish, but otherwise similar in its way of life.
Nycticebus coucang
Lorisidae

Senegal Bush Baby (72) differs from the loris by its long, bushy tail and large ear lobes. It weighs about 500 grams and measures about 40 cm including the tail. It is found in the savannas of central and southern Africa.
Galago senegalensis
Lorisidae

There are six species of bush baby in the savanna and tropical forests of southern, central and eastern Africa. Their diet consists mostly of insects; plants (fruits, juice, leaves) are just supplementary. They are arboreal, moving nimbly and speedily; they climb and leap well. Most species build nests of leaves, others use hollows in trees. They live in family groups, and like all prosimians, they communicate vocally. Two young are usually born; they are not carried around, but are left in the nest.

Potto (73) is the largest member of the family: it weighs up to 1.2 kg. The spines on cervical and thoracic vertebrae protrude and form a row of projections under the skin. This serves probably as a defence mechanism: when in danger the Potto rolls into a ball, turning its neck towards the enemy. It is strictly a nocturnal and arboreal animal inhabiting the middle layers of south-African rain forests. The single young is born well-developed, clutching to the mother's body immediately after birth and crawling over it.

Perodicticus potto
Lorisidae

Philippine Tarsier (74) represents the evolutionary line of Tarsioidea, indicating the course of evolution of primates, particularly with respect to the development of brain and dentition. Tarsiers are very conspicuous: they have a round head with large forward-directed 'owlish' eyes, and they can turn the head through 180°. They have short forelimbs, and very long hind limbs. The tips of the slender fingers are broadened and the nails are mostly flat; the thumb is opposed to the other fingers. The tail is long and totally naked except for a hairy tuft at the tip. Three species of tarsier occur in the Indo-Malayan island area, from Sumatra to the Philippines. They are confined to tropical jungles, particularly to coastal belts and thickets

Tarsius syrichta
Tarsiidae

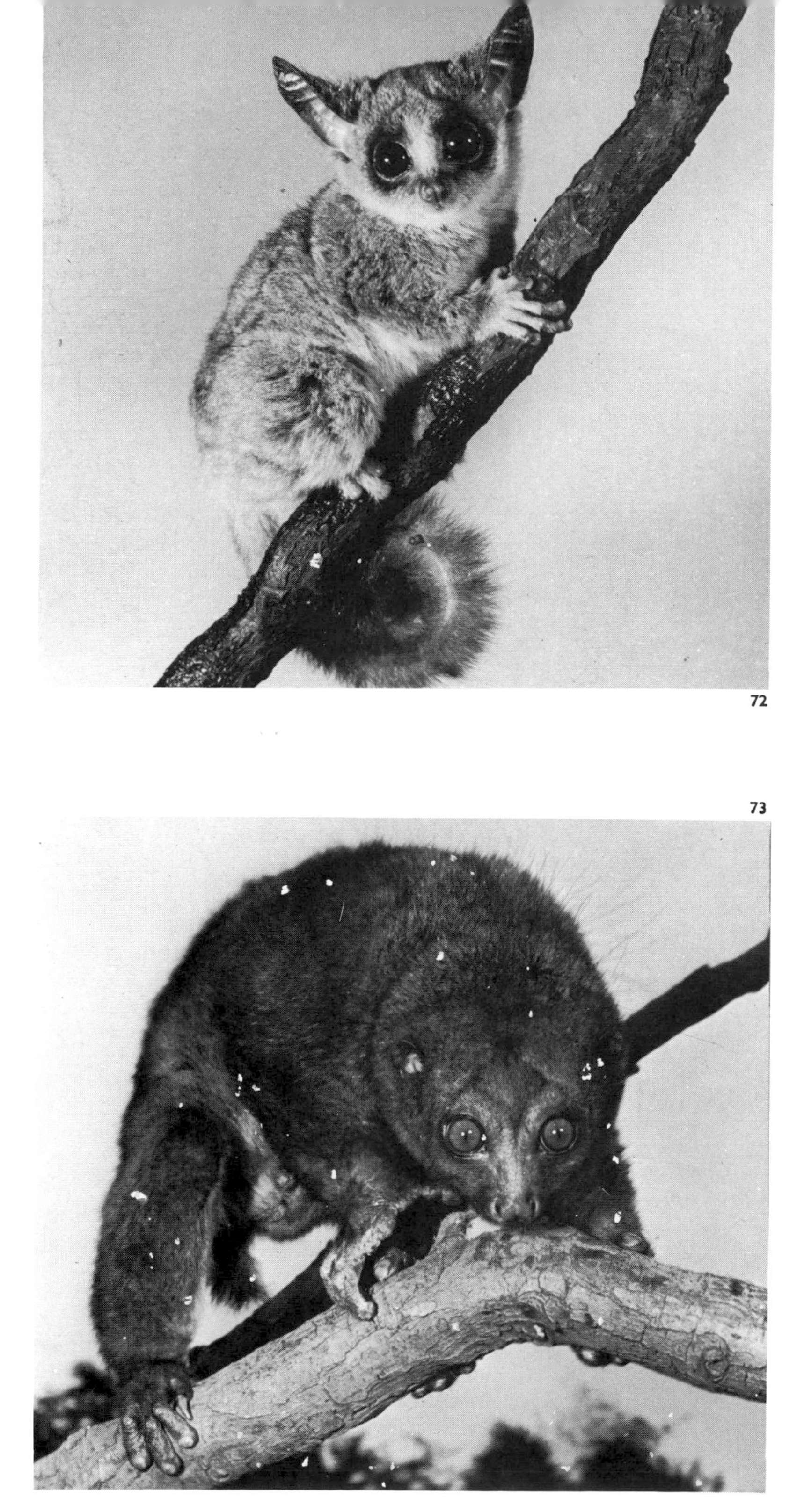

72

73

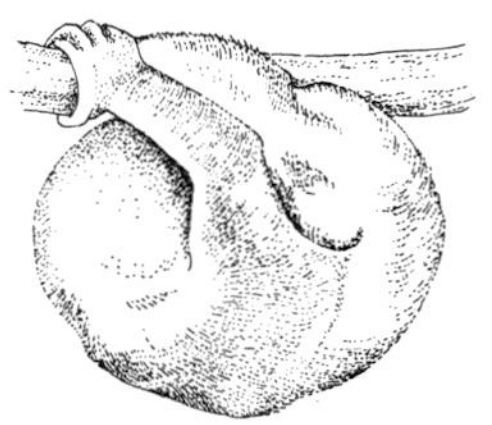

The Potto sleeps firmly attached to a tree branch

The Tarsier leaping

74

in river valleys, but sometimes occur even in plantations and gardens. They move on the ground by jumping like frogs, and are good climbers, too. They orientate themselves by sight and hearing and are active after sunset and early before dawn. They feed on insects and lizards, catch fish and crustaceans in brooks, and also consume fruits. They live singly or in pairs; females have a single young, which is born furry and is soon able to climb without help.

From an evolutionary point of view, New World monkeys are not of such importance as their Old World counterparts, but they demonstrate some of the interesting evolutionary potential of primates. There are about 70 classified species of New World monkeys; some of their populations have unfortunately been decimated by hunting and by destruction of habitats, and they are now among the most endangered mammals.

Marmosets and tamarins (Callithricidae) comprise about 33 species and form one of the two families of New World monkeys or the superfamily Ceboidea, which — in turn — represents one of the two large groups of higher primates of Anthropoidea. They are found only in the tropical virgin forests of South and Central America. Many of their significant morphological features support the conclusion that they developed independently during the lasting isolation of South America. They are descendants of prosimian ancestors, which had probably reached the South American continent in the Oligocene epoch. They are mostly smallish monkeys with soft, silky hair, naked cheeks and often with decorative manes. The nasal bone is broad, and their dentition consists of 36 teeth, whereas the Old World Anthropoidea, represented by the superfamily Cercopithecoidea, have only 32 teeth. Their limbs are also differently constructed. The thumb, which has no claw, is usually in line with the other, clawed fingers, not opposed to them. They have a long tail, used as a prehensile organ for gripping branches or grasping various objects. Most of them move in tree tops like squirrels. They are active by day and sleep at night in hollow trees. They gather in family bands consisting of a male, a female and several juveniles. They feed chiefly on insects, small vertebrates and sweet fruits. Females have one to three young, which reach sexual maturity in their second year.

Cotton-head Tamarin
Oedipomidas oedipus
Callithricidae

(75) is distinguishable from its numerous relatives by the typical tuft of long, white hairs on the forehead, white-coloured abdomen and limbs, relatively short ears and strongly pigmented genitals. It inhabits the damp zone of the Caribbean coast in Colombia and Panama.

75

Emperor Tamarin (76) represents the most numerous genus of American monkeys, including some 10 species of small and medium-sized monkeys with richly coloured coat and various arrangements of tufts of head hair. The Emperor Tamarin is easily recognized by its long 'imperial' moustache. It is found in north-western Brazil and is presumed to be present in adjacent areas of Peru between the rivers Jurua and Madre de Dios. It was discovered as late as in 1907 and received its name by mistake since taxidermists at first arranged its moustache in the style of that of the German Emperor Wilhelm.

Saguinus imperator
Callithricidae

Golden Lion Marmoset (77) is slightly larger than true marmosets and tamarins. It weighs 500–550 grams and measures 25–33 cm, with a tail 30–36 cm long. It has long, slender limbs with extremely long fingers. The 'lion' mane – longish hair on the neck and shoulders – gave it its name. It occurs in the coastal mountains southwards from Rio de Janeiro, at 500–1,000 metres of altitude. Two related lion marmosets, differing mainly in colour, live in restricted territories in Brazil. These are probably the remnants from an originally extensive distribution, now limited to a certain extent

Leontopithecus rosalia
Callithricidae

76

77

by the agricultural activities of man. Lion marmosets adapt with difficulty to changes accompanying civilization and disappear from cultivated regions: they are very rare, numbering among the most endangered South American mammals. They are arboreal, moving and leaping swiftly and safely in tree tops. Females give birth to a single young, which is born hairy and is immediately able to clutch at the mother's fur. The male also takes part in rearing the young.

Silvery Marmoset (78) belongs to the genus of small South American monkeys distinguished
Callithrix argentata
Callithricidae

mainly by boldly coloured tufts or crests of longish hair on the ears, by which the respective species can be identified. The species illustrated is widespread south of the Amazon between the rivers Tapjoz and Iriri, and in the south in Brazilian mountains. It is a larger species and is an exception to the rule, as its ears are without any tufts of longer hair. It measures 60 cm including the tail and weighs up to 350 grams. It lives both in trees of tropical rain forests and in the tall grass of savannas. It stays in groups and is active during the day. Silvery marmosets feed on sweet fruits, insects and smaller birds and their eggs.

Doroucouli (79) belongs to the other family of New World monkeys. It is the only truly nocturnal
Aotus trivirgatus
Cebidae

monkey, and daylight disagrees with it. It is active after sunset and at dawn, not in the dark. Its retina contains only rod cells and the monkey

78

cannot thus see in colour. It has clawed fingers and a long, furry, non-prehensile tail. The nasal bone is narrower than in other American monkeys. It is a rather small species in its family: its body measures 24—28 cm, the tail approximately the same. The Doroucouli moves in tree tops at a height of 6—30 metres, sleeps in tree hollows by day and never descends to the ground. It lives in pairs; the single young is nursed by the mother in the first days, then carried mostly by the father. It may reach 20 years of age. Its territory embraces Nicaragua, north-eastern Argentina, Guyana, Brazil, Peru and Ecuador.

Bald Uakari (80) has a purple, naked face and crown of the head. The face lacks subcutaneous fat pouches, which makes the monkey look scrawny and ill, but this is a false impression; according to observations made in the wild and in captivity, uakaris are extremely agile, climbing and jumping constantly. In their natural environment, they live in groups in tree tops, rarely coming down, and are active in the daytime. The diet comprises fruits,

Cacajao calvus
Cebidae

leaves, insects and small vertebrates. The Bald Uakari lives around the upper Amazon in Brazil and Peru. It is the only American monkey with a relatively short tail.

Sakis are typical residents of warm and humid rain forests in the heart of the South American continent. They are gregarious, living in tree tops at the edge of forests where food is abundant. They can jump adroitly from tree to tree, or walk erect. They feed on berries, nuts, leaves and smaller birds and mammals, particularly bats.

Pale-headed Saki (81)
Pithecia pithecia
Cebidae

is a rather large New World monkey (its body measures 37 cm, the tail 36 cm). The face is its outstanding feature: in males, it is covered by white-yellowish hair with black patterns around the eyes, nose and mouth and has a bare patch in the middle; in females, the face is black with a whitish stripe on both sides of the nose and on the chin. It lives north of the Amazon, as far as Guyana.

79

80

Hairy Saki (82) differs from the preceding species by its long, shaggy hair, partly covering the face like a wig and forming a sort of collar on the neck. It is widespread north of the Amazon, and farther west to Colombia and Guyana.

Pithecia monachus
Cebidae

81

82

The genus of American howler monkeys has been named after the powerful, howling call emitted by males (and sometimes females) when marking their territories. The sound is amplified by special resonators: bone drums, developed from the enlarged hyoid bone. Howler monkeys are arboreal; they move unhurriedly along branches and rarely leave the tree crowns. They have a long tail with a prehensile tip used as a fifth limb on which they can hang. They can make jumps as long as 4 metres. They live in groups of 15 to 20 members, led by senior males. Early mornings and late evenings are their active time. They feed on soft fruits, leaves, shoots, and occasionally on insects and small vertebrates. The female produces a single young which immediately clutches to the mother's furry abdomen, and after a month moves to her back. The baby is suckled for 12 months and is cared for by the whole group. Howler monkeys are widespread in a large territory from Vera Cruz in Mexico, south of Ecuador, and from the Andes across Bolivia, Brazil and Paraguay.

83

Red Howler Monkey (83) has reddish brown fur (related species are mostly black), longer on the back and head. It is found in the virgin forests of central South America, in Venezuela, Colombia, Ecuador, eastern Peru and western Brazil. Together with five related species, it is one of the largest South American monkeys: its body measures about 60 cm, and the tail is just a little shorter.
Alouatta seniculus
Cebidae

Brown Capuchin (84) is about 45–55 cm long and weighs about 2–2.5 kg. There is the suggestion of a flat cap on its forehead. The long and furry tail is black at the tip and only slightly prehensile; when walking, the Capuchin carries it curled up. It lives in the forest regions of South America from Colombia and Venezuela to Argentina.
Cebus apella
Cebidae

All four species of capuchin belong with the most abundant and typical representatives of New World monkeys: they are widely distributed over the South and Central Americas from Costa Rica to Paraguay, absent only from the north-west of the continent. They inhabit warm and humid primary forests in both the Amazon plains and the foothills of the Andes. They are arboreal, living in family groups or larger parties, and visiting the ground only to drink. They are active in the daytime, feeding on pulpy fruits, shoots, leaves, insects and small vertebrates. A single young is born, after 6 months of gestation.

Black Spider Monkey (85) has all the features typical of its genus: long, slender limbs, shortened body, a relatively small head and particularly a long and mobile, prehensile tail – the most perfected tail of all monkeys. The underside of its tip is naked, with touch-sensitive strips, and it serves as a full fifth 'limb' not only in movement, but also when handling food or investigating various objects. Spider monkeys are comfortably at home in tree tops. They climb dexterously and before jumping, they swing on their tails, thus attaining distances up to 7.5 metres. On the other hand, they
Ateles paniscus
Cebidae

can walk on their hind limbs — an unusual feat for American monkeys. They live in small parties of 15 to 20 members, with a poorly established hierarchy: males only rarely fight, and then only in the absence of females. Spider monkeys feed chiefly on fruit and oleaginous plant parts. Their incisors are comparatively powerful. Spider monkeys always have a single young which is carried at first on the abdomen, later on the back. The Black Spider Monkey is usually dark or black-coloured, with a dark or pale face; the colouration differs considerably in individual subspecies. It occurs in the north of South America, reaching from the upper Amazon region up to Central America. It is commonly used as a laboratory animal for malaria research.

Common Squirrel Monkey
Saimiri sciureus
Cebidae

(86) has a characteristically patterned face with a dark spot on the muzzle. It is one of the smallest New World monkeys, measuring only 22—30 cm, with a tail of about the same length. Squirrel monkeys are widely distributed in South and Central American virgin forests, and at their edges or in forest galleries. They are very abundant and live in small groups in trees, feeding chiefly on fruits but consuming also large quantities of insects and invertebrates. They are active by day. Although having a ruling hierarchy, they rarely fight; disputes are settled by conciliatory behaviour. Reproduction is not limited to a specific season; a single young is born and is carried around by one of the parents. In proportion by weight, squirrel monkeys have the largest brain among primates — 1/17 of the body weight — but it is only slightly convoluted. It is estimated that the most developed regions of the brain are those controlling orientation in complicated movements, particularly in jumps.

84

The Spider Monkey uses its tail in climbing

85

Black-and-White Colobus (87) is one of some 60 species belonging to the superfamily of Cercopithecoidea, native to the Old World, and differing from American monkeys in having a narrow nasal bone and only 32 teeth. They have either a longish, non-prehensile tail, or a shortened one. Nails are present on all fingers, but the thumb is often reduced. Originally arboreal, many species secondarily adopted a ground-living terrestrial way of life. Colobus is considered to be one of the most beautiful African monkeys. It is distinguished from the other monkey species above all by the lack of a thumb. It is the only completely vegetarian African monkey, feeding on leaves of certain trees. It lives in forests and virgin forests from lowlands high up into the mountains. Small groups of 20 individuals are led by one male; their territory covers an area of roughly 10 hectares and these are marked by sound and visual signals. Females bear a single young, which is born white and only later develops adult colouration, which is highly variable. Colobus monkeys are strictly protected nowadays. The Black-and-White Colobus is widespread in Africa from Senegal to Ethiopia, and from Angola over Zaire to Tanzania.

Colobus polycomos
Cercopithecidae

Capped Langur (88) is closely related to colobus monkeys, but it is confined to Asia: Assam, Burma and Yunnan. It is a relatively large monkey, measuring 50–65 cm excluding the tail.
Presbytis pileata
Cercopithecidae

Dusky Leaf Monkey (89) is recognized by the conspicuous white eye pattern, reminiscent of spectacles. It is native to the humid forests of Tenasserim, Indo-China, Thailand and the Malayan Peninsula. Like all leaf monkeys (there are 14 species), it lives in small groups and feeds on plant food. It is a good climber and moves quickly even on the ground.
Presbytis obscura
Cercopithecidae

Proboscis Monkey (90, 91) received its name because of the huge, bulbous, overhanging nose of old males. The nose entirely covers the mouth and the monkey pushes it away when eating. It functions as a resonator of sounds by which the males denote their territories. Females and young have a short, turned-up nose. The Proboscis Monkey inhabits the primary forests of Borneo (Kalimantan). It measures 66–75 cm, and the tail is equally long. It is a very rare species nowadays. Its way of life is little known; it is arboreal, living in small groups, most frequently in mangroves or near rivers. Its diet consists of leaves and fruits.
Nasalis larvatus
Cercopithecidae

Douc Langur (92) is one of the two species of the peculiar genus of langurs, characterized by their noses and conspicuous colouring. It lives in Vietnam, in southern Laos, and in view of its rarity, it has been listed as one of the endangered species. In its way of life, it hardly differs from leaf monkeys and it is closely related to them.
Pygathrix nemaeus
Cercopithecidae

86

Rhesus Macaque (93) *Macaca mulatta* Cercopithecidae

is the best known member of the genus of macaques, medium-sized monkeys with round faces and prominent supraorbital ridges in males. It is widespread in southern Afghanistan, India and southern China. It is a common monkey of zoological gardens and of laboratories, where it is reputed as a good experimental animal. For instance, the so-called Rh (Rhesus) factor, a hereditary quality of blood proteins, which in certain combinations can be fatal for newborn human babies, was discovered in rhesus monkeys; also, these were the first monkeys to take part in space exploration. Unfortunately, the vast exploitation of rhesus monkeys in scientific research, and huge losses caused by unsuitable transport, are decimating this formerly abundant animal so that serious steps have to be taken for its protection. In some areas, they are protected for religious reasons; elsewhere they are hunted as pests. Rhesus monkeys live in forests, open rocky lowlands and cultivated regions. They feed on various fruits, shoots, leaves, roots and smaller animals. They live in groups of 50 individuals at the most. The females give birth to a single young which is suckled for almost a year.

88

89

90

91

92

93

Japanese Macaque (94) has typical naked and bright red cheeks, edged by a long mane. It is a relatively large species with a rather short tail and a body length of 55–70 cm; its weight is 8–12 kg. It is the northernmost not only of all macaques but also of all non-human primates: it lives on the Japanese islands, and plays an important part in Japanese mythology, literature and art. Macaques include about 10 other species, inhabiting southern Asia from Afghanistan to the Philippines, Taiwan and Japan, and north-western Africa. They even include the Barbary Ape (*Macaca sylvanus*): since it occurs in Gibraltar, it is the only primate (except Man) extending to Europe. The Crab-eating Macaque (*M. fascicularis*), a little smaller than the rhesus monkey, is nowadays imported in quantity from Indonesia as an experimental animal. The Lion-tailed Macaque (*M. silenus*) from southern India is the rarest species.

Macaca fuscata
Cercopithecidae

Gelada Baboon (95) is recognized by the red, bald spot on the chest, noticeable mainly in the breeding season, and by the pink-rimmed eyes. The males are distinguished by a cape of longish hair and whiskers. Adult males reach up to 70 cm in body length; the tail is about 50 cm long. The Gelada Baboon lives in Ethiopian mountains at altitudes of 1,800–4,000 metres, in rocks

Theropithecus gelada
Cercopithecidae

94

and bushes. It is chiefly terrestrial and is active during the day; at night, it rests in caves and shelters among rocks. It forms large, probably family groups, led by the old males. In the event of danger, gelada baboons produce barking sounds and even throw stones at the intruders. They eat plant food.

Anubis Baboon (96) is a representative of the true baboons, stocky monkeys with elongated, 'dog-like' muzzles, powerful canine teeth, prominent supraorbital ridges and well-developed seat calluses. It is widespread in eastern Africa, from Ethiopia to Kenya and Tanzania, westwards to Nigeria, and in places in the Nigerian Ennedi and Tibesti Mountains. Its favourite haunts are steppes and savannas; it is so numerous in places that it causes serious damage to plantations. Packs of 40 to 80 (200 at most) have a complex

Papio anubis
Cercopithecidae

Distribution of baboons in Africa

95

hierarchy. Baboons are active in daytime, at night resting in trees, caves and rocks. They walk on all fours, with the tail carried arched in an inverted 'U'. The female bears one, rarely two young, carrying them on her chest and later on her back, and suckling them for 5 to 6 months. The diet comprises fruit, leaves, grass, roots, bulbs, and occasionally reptiles, birds' eggs or, exceptionally, young mammals up to the size of a gazelle.

96

97

98

Mandrill (97, 98) is the largest baboon. Males weigh up to 45 kg, thus reaching almost the size of some apes. Females are smaller and much less colourful. They live in the coastal virgin forests of western Africa in Cameroon, Guinea and Zaire. Some of their features, for example the powerful canines and the large thumb, which developed as adaptation to life in the forest, led some zoologists to separate this and the next species, into the individual genus *Mandrillus*. Mandrills are the most highly coloured of monkeys and mammals in general. Bright colours on the face, particularly striking when the animals are excited, enhance the menacing attitude of wide-opened mouth and bared teeth. These displays are used against enemies and intruders and against inferior members of the group. In contrast, showing the colourful rump is an expression of passivity and submission.

Papio sphinx
Cercopithecidae

99

100

Drill (99) is another forest baboon, inhabiting roughly the same territory as the preceding species,
Papio leucophaeus
Cercopithecidae

but also Nigeria and the island of Fernando Po. It is somewhat smaller than the Mandrill, weighing up to 30 kg. The face is black, without prominent cheek bones, only the chin is red and the whole face rimmed by a whitish beard. Drills feed on fruit, berries, bark, shoots, molluscs, worms, insects, reptiles and amphibians, rarely on mammals. Large packs of up to 60 individuals usually split up into several smaller groups. The single young is carried on the mother's back.

Sooty Mangabey (100)
Cercocebus torquatus
Cercopithecidae

represents the genus of mangabeys, close relatives of baboons and macaques. In appearance, way of life and distribution, however, it is more reminiscent of monkeys of the genus *Cercopithecus*. The four known species of mangabey are mostly arboreal animals. They inhabit damp, swampy forests and river banks but also cultivated environments. They are quite abundant and live in packs, feeding on plant food and smaller animals. The single young is carried by the mother on her abdomen. The Sooty Mangabey is a slender monkey with a short, white-rimmed face, reddish-brown crown and a long tail. It is found in western Africa from Cape Verde to the lower part of the Zaire (Congo) river.

101

De Brazza's Monkey *Cercopithecus neglectus* Cercopithecidae

(101) is one of the many species of the genus *Cercopithecus,* widespread in Africa south of the Sahara. The systematic classification of this heterogenous group has not yet been firmly established: it lists some 14 to 20 species and up to 70 subspecies. All of them are slender, small monkeys; they have a rounded head, flat face, inconspicuous nose with a narrow nasal bone, and usually a long tail. They are olive-coloured, greyish to black on the back and paler on the belly. The neck, flanks, head and rump

102

103

often bear conspicuous markings. Certain typical features distinguish them from the other Old World monkeys: for example molars with only four cusps, imperceptible seat calluses and traces of prehension of the tail in young monkeys. They inhabit forests and savannas; in open landscapes, they concentrate in wood galleries near water and never occur in arid places. De Brazza's Monkey has a dark coat, striking white beard, reddish forehead and white stripes on the thighs. Adult animals measure 45–60 cm and weigh up to 8 kg; the tail is about 70 cm long. It is found in the vast area from Cameroon to Kenya and Angola. It inhabits both lowlands and mountains and, more frequently than other related species, moves on the ground. Insects are an important part of its diet.

Diana Monkey (102)
Cercopithecus diana
Cercopithecidae

is the most variegated cercopithecus. Its colouration includes shades of black, white, brown and yellow-orange; furthermore, males have a remarkable white beard. Diana Monkeys occur in the humid, coastal, virgin forests of western Africa from Sierra Leone to Ghana and Zaire. More than any other monkey species, they are restricted to trees and feed almost exclusively on their fruits. They are active during the day, living in groups of several or several hundred individuals.

Red Guenon (103)
Erythrocebus patas
Cercopithecidae

is mostly terrestrial and climbs trees only in case of danger. This is reflected in certain features of its anatomy; for instance, its legs are adapted for running, being very long with short toes. Males weigh up to 8–12 kg. They are pale rusty brown with whitish underparts and a blue-pigmented sexual organ; older guenons have white whiskers and a mane. Females are less colourful and half the size. Guenons live in grassy savannas with acacia woods, usually in groups of seven to 15, led by the senior male.

They are active during the day, feeding on insects, fruits, leaves, shoots, and occasionally tiny vertebrates and eggs. A single young is born. They are distributed throughout central Africa, from Senegal to Sudan and Ethiopia, and south to Cameroon, Zaire and Tanzania.

The family of gibbons belongs, with apes and Man, to the superior group of primates (superfamily Hominoidea), and therefore have more features in common with apes than with monkeys. These features include above all a relatively large brain, broad chest, flat face with shortened jaws, shortened lumbar section of the spinal cord, and stunted tail. However, other features, developed under the influence of arboreal life — disproportionally elongated and mobile front limbs with slender, long-fingered hands and a separated thumb — now dominate the gibbon's appearance. The hair on the arms grows downwards from the shoulder to the elbow and upwards on the forearm. Gibbons keep to the treetops where they move with agility. Their main manner of locomotion, called brachiation, involves the arms more than the legs: they hang on branches, swing, and easily overcome distances of up to 10 metres by leaping. Their fast and accurate movement through the branches resembles running. They can move adroitly on the ground, walking erect on hind legs and supporting themselves with their hands. Gibbons, like

105

apes, cannot swim and avoid water; they even drink very little. They feed mostly on plant food and usually gather in small family groups. Generally, gibbons are sexually mature in the seventh year. Following 210 days of gestation, the female produces a single young and carries it on the abdomen for several months. Sounds, produced by their deep, baying voices, surpass all the other animal sounds. They are amplified by highly developed throat pouches and can be heard in the morning and evening; in this way individual packs denote their territories. Gibbons are found in south-eastern Asia and in the Indo-Malayan archipelago.

White-handed Gibbon (104) is brown-yellow or brown-black, with a naked dark- and white-lined face. It stands 40 – 50 cm and weighs 4 – 13 kg. It inhabits humid tropical forests of Indo-China and Sumatra.

Hylobates lar
Hylobatidae

Siamang (105) is the largest gibbon, reaching up to 95 cm when standing and weighing about 18 to 22 kg. It differs from other gibbons by the long hair, the enlarged, bare throat patch and particularly the fused second and third toes of the hind (sometimes also front) feet. Consequently, it is classified as a separate genus. In comparison with other gibbons, it is said to move more slowly and clumsily. The extraordinarily strong voice is produced by both sexes. The Siamang lives in the jungles and mountainous forests of the Malay Peninsula and in Sumatra.

Symphalangus syndactylus
Hylobatidae

Orang-utan (64, 106) belongs, with the Gorilla and Chimpanzee, to the family of apes (Pongidae), characterized by a large cranium and well-developed brain, broad chest and pelvis, elongated arms and relatively short legs, highly developed muscles and lack of a tail. Most of these features are common with men, which testifies to the close relationship of apes with the family Hominidae, even though it is obvious that apes were not the direct ancestors of Man. The Orang-utan is the largest species after the Gorilla: the female weighs

Pongo pygmaeus
Pongidae

35—40 kg, the male up to 80 and exceptionally even over 100 kg. The standing height of old males is 135—150 cm. The long, reddish coat sparsely covers the blue-black skin. The face is naked and dark, in old males with marked rounded cheek callosities and a naked throat pouch shaped like a large goitre. It has, relatively, the longest arms of all apes, reaching 2.25 metres when stretched. The last survivors (some 2,500 specimens) live in the tropical rain forests of Borneo (Kalimantan) and Sumatra and their future is uncertain, despite strict protection. Orang-utans are arboreal and active by day, feeding on fruits (mostly figs and durian), leaves, shoots, seeds, but also on birds and eggs. They sleep in tree nests of branches and leaves usually built at a different site from the one used the preceding night. They are solitary or live in pairs and small family groups. After 8—9 months of gestation, the female bears a single young

weighing 1.5 kg at birth. The mother stays with it until it is four or five and it reaches maturity when 10—12 years old. Life expectancy of Orang-utans is 30—40 years.

Gorilla (107) *Gorilla gorilla* Pongidae

is the largest of apes: females weigh 70—125 kg, males 130—218 kg, in captivity even 300—350 kg. Standing males measure 1.8—2 metres and the span of their stretched arms reaches up to 2.75 metres. The Gorilla has a large head with prominent crown, flat face, low, wide nose and short ears. The body is covered by thin black hair with a silvery 'saddle' on the back in old males. The forelimbs are stocky and elongated, the hind limbs are relatively short. The Gorilla walks on the soles of its hind feet and supports itself with the bent fingers of the hands (the so-called 'knuckle-walking'). It inhabits tropical forests and bamboo thickets of equatorial Africa, especially lowlands of Zaire (Congo) and western Africa (Lowland Gorilla, *G. g. gorilla*) and the mountains of eastern Zaire and western Uganda (Mountain Gorilla, *G. g. beringei*). Its numbers have been considerably reduced in recent years: it is estimated that about 5,000 gorillas still survive; they are among the endangered species. Gorillas are active by day and they are chiefly terrestrial. They gather in packs of five to 30 individuals, composed of one dominant and several subordinate males, and an approximately double number of females and juveniles. They spend the night sleeping in nests of branches, mostly

107

108

built anew every evening, in grass, among rocks or in trees. Thea eat mainly leaves, bark, bamboo shoots, roots and fruit. Once in about 4 years, females bear a single young, weighing 2 kg at birth and growing slowly to reach maturity at the age of 8–12 years.

Chimpanzee (108), with its naked face and large ears, gesticulations and mimicry resembles Man most of all the apes. A standing adult male measures 1.1–1.3 metres and weighs 45–80 kg; females are smaller. The body is covered by thin, black hair; in older chimps, the chin becomes grey or bald. The powerful, mobile limbs have long big toes on the feet, and shortened thumbs on the hands. The limited number of chimpanzees live in central Africa, east of the rivers Zaire and Niger, and they are protected. They inhabit tropical forests and they are both arboreal and terrestrial. They are active by day, travelling in packs of 2–20 individuals, led by a dominant male. They orientate themselves by sight and hearing. Having the superior mental capacities of all apes, they are kept in research institutes as experimental animals.

Pan troglodytes
Pongidae

109

Chapter 6 ANTEATERS AND THEIR ALLIES

Classification of a species of mammal into its respective order might best be determined by the animal's teeth: their number, variety and structure; but there are mammals in which this significant feature is irrelevant and may lead to error. Such is the case with the group of mammals described in this chapter. In the last century, three different groups were classified into the order of edentates (Edentata); as it turned out later, the three were not related. The three groups comprised pangolins, aardvarks, anteaters, sloths and armadillos. This classification was based on their strange dentition: as a result of extreme specialization of feeding, the characteristic features were progressively simplified. All three feed only on insects, occasionally on tender parts of plants. Modern understanding includes only anteaters (Myrmecophagidae), sloths (Bradypodidae) and armadillos (Dasypodidae) among the order Edentata. The name edentates, or toothless, is in fact incorrect, for only anteaters are totally toothless, both sloths and armadillos possess some simplified teeth: the largest armadillo has an outstanding number of teeth for a mammal: about one-hundred. Pangolins, which represent a different order, Pholidota, are also toothless, but this is merely a consequence of their food specialization, and it does not prove any relationship to the true edentates. Similarly, simplified teeth in African Aardvarks, now classified into the order Tubulidentata, are the result of specialization for a diet of termites. Taking the dentition as the only proof of a relationship will thus lead to wrong conclusions. The true edentates are an ancient group of mammals; their origin (the early Tertiary) and further evolution are confined to the New World, particularly to the once isolated South American continent, where some of them still survive. Only recently were the territories of some species extended northwards, to Central America and to southern parts of the North American continent.

110

111

112

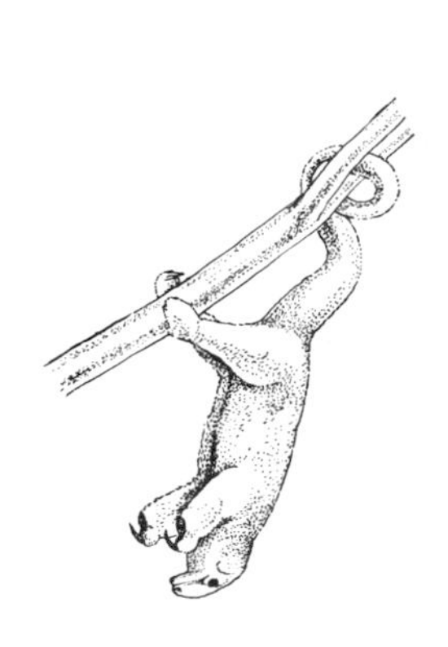

The Two-toed Anteater *(Cyclopes didactylus)* is the smallest and rarest of the three known species of anteaters

113

Great Anteater (109, 110) is one of the three species of the family of anteaters. Thanks to its characteristic appearance, it is easily recognized even by a layman. Including the tail, it measures about 2 metres and weighs 18–23 kg. In proportion to the body, the tubular head of the Great Anteater is small, with small ears and a narrow, pointed snout terminating in a narrow mouth. The forelimbs have strong, sharp claws; when the animal walks, they bend backwards so that it walks on the backs of its fingers. The body is covered by long, coarse hair creating a mane on the back and long tufts in the tail. Anteaters dig out their food – chiefly ants and termites – from anthills and termitaries, sticking them to the long, pencil-like and sticky tongue, which can be drawn out as far as 60 cm. There are no teeth in the jaws; the food is crushed in the strong-walled stomach. About 30,000 termites may be consumed daily. The Great Anteater is widespread in the swampy and humid rain forests and savanna of Central and South America from Honduras to northern Argentina. It is active mostly during the day and it is a good digger and swimmer. It leads a solitary life, inhabiting abandoned burrows of other animals, or rocky clefts. The female bears a single young, which she carries on her back.

Myrmecophaga tridactyla
Myrmecophagidae

114

Tamandua (111) measures about 1 m from head to tail. It has short hair, a short head, and is
Tamandua tetradactyla
Myrmecophagidae

more conspicuously coloured than the preceding species. It is a nocturnal, arboreal animal, and it climbs nimbly thanks to its sharp claws and a partly prehensile tail. It also finds its food on trees; mostly ants and termites living in the wood. It inhabits southern Mexico, Bolivia and Peru.

Two-toed Sloth (112) is the most abundant species of the family Bradypodidae. It lives mostly
Choloepus didactylus
Bradypodidae

in trees, moving slowly and hesitantly and feeding on leaves, young shoots, twigs and fruits. Its fur is thick and waterproof; in the rainy season, green algae multiply in it to such an extent that it acquires a protective greenish colouring. Sloths are confined to the primary forests of South and Central America from Honduras to northern Argentina, Paraguay and Brazil.

Armadillos (Dasypodidae) are the only mammals in which the skin developed into a special bony and horny carapace in the form of bands of plates connected by flexible skin and covering the back, head, sides and tail. The number of free-moving bands on the back differs in various species, and therefore serves as an aid to identification. Unlike other edentates, armadillos have numerous tiny and peg-like teeth (seven to 25 in each jaw); the teeth grow constantly. These terrestrial animals are good diggers, and feed on insects, tender parts of plants, small vertebrates and carcasses. The females may produce several young from one fertilized egg, which divides into several embryos (twelve at the most). The young are thus always of the same sex. This phenomenon is called polyembryony; it occurs only occasionally in other mammals. Armadillos also sometimes have prolonged gestation and variable body temperature. There are about 21 species, distributed from the southern United States to Patagonia. The largest, *Priodontes giganteus*, weighs up to 55 kg; the smallest species of the genera *Burmeisteria* and *Chlamyphorus* weigh a mere 90 grams.

Hairy Armadillo (113)
Euphractus villosus
Dasypodidae

has seven or eight bands of bony plates with tough bristles scattered between them. It is a medium-sized species weighing 2.5—3 kg. It lives in arid savannas of Argentine, Peru and Chile. The Hairy Armadillo is a nocturnal and mostly vegetarian animal. It is abundant in places and may cause damage to crops.

115

Apara (114) has a flexible carapace, which permits it to curl into a ball when attacked. The three inner toes of the hind feet are fused together. The Apara inhabits the open grasslands, swampy savanna and humid forests of the north-western part of central Brazil. It feeds mostly on ants and termites. Unlike other species, females produce only a single young. The Apara is relatively small, measuring some 35–45 cm.

Tolypeutus tricinctus
Dasypodidae

Nine-banded Armadillo (115) lives farthest north of all armadillos, reaching Mexico and the southern United States. It also inhabits the whole of Central America and large parts of South America east of the Andes and south to northern Argentina and Uruguay. It can be identified by the nine back bands, long ears and a relatively long tail; it weighs about 6 kg. It lives gregariously in self-dug burrows. It is mostly nocturnal and feeds on insects, worms, molluscs, small vertebrates (sometimes it attacks venomous snakes), soft fruits, seeds and mushrooms. One litter averages four young; life expectancy is 4 years.

Dasypus novemcinctus
Dasypodidae

Chinese Pangolin (116) is one of the seven living species of pangolins (Pholidota): their representatives are among the most extraordinary mammals on Earth. Their body cover is their most remarkable feature: including the tail, they are covered by thick, horny scales, overlapping like scales on a spruce cone. The only uncovered parts are the abdomen, throat, muzzle, cheeks and insides of the limbs; these are sparsely covered with tough bristles. Pangolins have small, slender heads with tiny eyes and without noticeable external ears. The muzzle has a narrow mouth; the teeth are completely missing but the long, worm-like and extensible tongue, situated in a special tongue sheath, is considerably developed. Its movement is made possible by the contraction of muscles attached to an elongated protuberance of the sternum. The sticky tongue traps ants and termites, the pangolin's staple food. Insects are crushed in a special stomach compartment with muscular walls and hard, horny projections from the mucuous membrane. Pangolins live in Africa south of the Sahara, and in south-eastern Asia from India to southern China, Taiwan and Indonesia.

Manis pentadactyla
Manidae

Tree pangolins use their prehensile tails in climbing

116

117

They keep to forests and bushy savanna; some species are terrestrial, others arboreal. They live solitarily or in pairs, sleeping during the day in underground burrows or tree hollows. There are one to three young in a litter. The Chinese Pangolin is a medium-sized animal; it measures 50—60 cm and has a comparatively short tail. It is widespread in Taiwan, southern China and Indochina.

Aardvark (117)
Orycteropus afer
Orycteropodidae

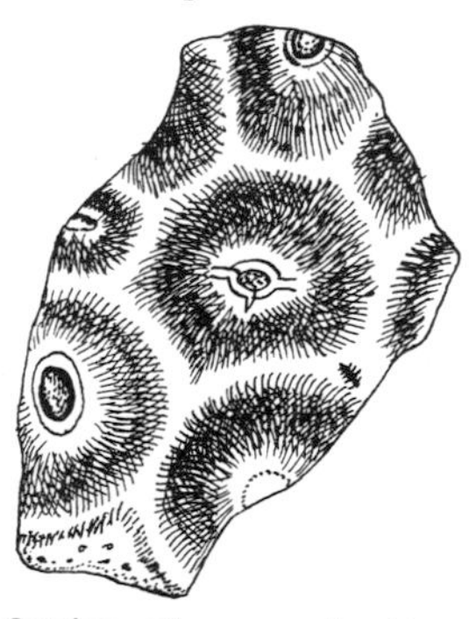

Section of an aardvark's tooth showing its unique structure

was originally considered to be a member of the order of edentates. It has been proved, however, that it is related to ungulates and is the only surviving member of the order Tubulidentata. It has interesting teeth: eight to ten peg-like, rootless teeth composed only of dentine and cement, built of hexagonal columns with central canals for nerves and capillaries. Milk dentition comprises several types of teeth, and these helped to determine the true origin of aardvarks. Externally, the group do not resemble ungulates at all. They are relatively large (measuring about 2 metres including the tail, and weighing 70—80 kg), they have a stout body with an arched back, a long head with a pig-like muzzle, large eyes, cornet-like ears and a thick tail. Their skin is mostly naked, except some sparse bristles on the muzzle, sides and limbs. They feed almost exclusively on insects, chiefly termites, and only exceptionally on fruits. They live throughout Africa south of the Sahara in various environments, except the forests. These nocturnal animals spend the day sleeping in deep burrows, which they dig themselves. They are solitary and secretive and their life style is little known.

118

Chapter 7 RODENTS

The success of each mammalian order can be considered from various viewpoints; if the number of contemporary living species were taken as the criterion, rodents (Rodentia) would pass with flying colours. There are about 1,700 classified species, which represents almost two-fifths of all known mammalian species. Rodents have settled all over the world, making an important contribution to the fauna of areas in which even vegetation is scarce. Their successful spread has been assisted by numerous useful adaptations, particularly their specialization on plant food, which is found in abundance almost everywhere. Rodents are able to consume even such parts of plants as are indigestible to other herbivores, for example roots, bark, tough grass or wood. Their small size is another vital factor: among other things, the period required for embryonic development is short, replacement of generations rapid, and the reproductive potential is immense.

Rodents are an ancient group with ancestors from the very beginning of the Tertiary. Their differentiation into a variety of evolutionary lines took place at an early stage; contemporary forms, differing in appearance and ecology, developed from them. Unrelated species are often similar in appearance and way of life as a result of parallel evolution. Specialists often find it difficult to differentiate and assess correctly those features of adaptation gained under the influence of a specific way of life, and those characteristics which are phylogenetic (ancestral); this complicates the final arrangement of the classification of rodents.

Members of the large suborder Sciuromorpha typically have simple masticating muscles and more than three cheek-teeth on each side of the lower jaw. They include eight rather different families. The family of squirrels (Sciuridae), with over 225 recent species, is the most typical of them.

Red Squirrel (118) is a typical example of a rodent adapted to arboreal life. When climbing, its sharp claws hook into the bark; it can leap from tree to tree, slowing its descent with widespread limbs and long, bushy tail. It lives in forests, particularly coniferous ones. Squirrels feed on the seeds of trees (mostly

Sciurus vulgaris
Sciuridae

119

of conifers), skilfully picking them out of the cones; they also gather nuts, berries, fruit, young shoots, mushrooms and occasionally insects, birds' eggs and fledglings. In autumn, they make stores of food in tree hollows. They are active during both winter and summer; only a short period of the coldest winter is spent in hollows or in round nests of twigs and leaves built in trees. Females have two litters of three to seven young a year; in summer, abundance of food can provoke overpopulation. Two colour variants exist: red, and black-brown; the abdomen is always whitish. Squirrels are widespread almost all over Europe and in the forest regions of Asia as far east as Japan.

Grey Squirrel (119) inhabits deciduous forests, gardens and parks of eastern Canada and the United States. It was introduced into the British Isles. The Grey Squirrel is larger than its European counterpart; it lacks ear tufts and is mostly greyish.
Sciurus carolinensis
Sciuridae

Arctic Ground Squirrel (120) lives in Canada, Alaska, islands of the Bering Strait, and northern Siberia. It is one of the 20 species of a genus widespread in steppes, forest steppes and forest tundras of the Eurasian and North American temperate zone.
Citellus parryi
Sciuridae

California Ground Squirrel (121) lives in short grasslands, in the western and south-western United States and the Mexican part of the Californian peninsula. In winter and in the driest summers, it sleeps in burrows which it digs itself.
Citellus beechei
Sciuridae

European Souslik (122) is one of the so-called ground squirrels. Originally a steppe-dweller, it now inhabits all central and south-eastern Europe, as far as western Ukraine and Moldavia. In agricultural regions, colonies of sousliks settle in uncultivated fields, arid pastureland, edges of fields and roads, and
Citellus citellus
Sciuridae

120

121

railway banks. They dig complex networks of underground corridors as much as 2 metres deep. They are mostly active in the morning and early evening. When in danger, they emit a warning whistle and retreat rapidly underground. Hibernation lasts from late summer to the end of March or April. During that time they live on the fat stored in the body during the summer. The single annual litter averages six to eight young. In favourable years overpopulation may occur.

122

123

Alpine Marmot (123) represents the genus of marmots, including some 13 species widespread in Canada, the United States, and in Eurasia from the Alps to north-eastern China, the Himalayas and Yunnan. It is found in its original habitat only in the Alps and High Tatras; but it was successfully introduced to the Pyrenees, Low Tatras and Schwarzwald. The total length

Marmota marmota
Sciuridae

124

125

is 60—73 cm, and the weight 5—6 kg. It lives gregariously, on grassy and stony slopes of the Alpine mountain regions, most often at 1,300 to 2,700 metres of altitude. Its diet consists of green parts of plants. It is a diurnal animal. Its burrows, as much as 10 metres long, reach a depth of 3 metres and have several exits. When danger threatens, the Alpine Marmot emits a sharp warning whistle. From September to the end of April or May, marmots hibernate deep down in their grass-lined burrows, living off body fat stores. In late May or early June, females bear two to six naked and blind young. Alpine Marmots are strictly protected, and their numbers have increased in recent years.

Bobac Marmot (124)
Marmota bobak
Sciuridae

unlike the majority of marmots, is a steppe-dweller. It inhabits the lowland and upland steppes of eastern Europe and western Asia. It used to be abundant in the steppes of the Ukraine as well as those of Moldavia and Rumania, but it was almost eliminated in the western part of this area as a result of the spread of land cultivation and its being hunted for its fur. It is the same size as the Alpine Marmot; its coat is yellowish with a darker head and a paler abdomen. It is probably the original form out of which the mountain species developed. It is prone to plague, which can affect men; the disease is spread by fleas parasitizing on the animal's body.

126

Woodchuck (125) is a resident of the open flatlands of North America, from Alaska and western Canada to the U.S. Atlantic coast. In its way of life, it does not differ from other marmots.
Marmota monax
Sciuridae

Black-tailed Prairie Dog (126) weighs about 700 – 1,200 grams. It is named from its call, which resembles a dog's barking. It is an original and formerly very abundant inhabitant of North American prairies. It is a gregarious animal; it digs intricate burrow networks, composed of vertical entrance corridors and horizontal connecting passages with nesting chambers. Numerous exits on the surface are protected from rain by round mounds of excavated earth, and are linked by tracks.
Cynomys ludovicianus
Sciuridae

127

The Southern Flying Squirrel in a gliding jump

128

Burunduk (127) is closely allied to sousliks, but it is better equipped for tree-climbing: it has sharp, sickle-shaped claws and soft pads on the paws. It is a diurnal, lively animal widespread in the northern forests from the easternmost tip of Europe, across Siberia, to China and Japan. Thickets, forest borders and riverside shrubberies are its favourite haunts. The Burunduk feeds on seeds, berries, insects and sometimes on small vertebrates. It hibernates from October to April in underground nests, waking up from time to time and feeding on stored food.

Eutamias sibiricus
Sciuridae

129

130

Eastern Chipmunk (128) is recognized by the light stripe on the flanks and head. It is diurnal and moves on the ground with its tail held vertically. It has a characteristic voice. Northern populations hibernate. It is distributed in the eastern parts of North America.
Tamias striatus
Sciuridae

Southern Flying Squirrel (129) measures only 12–13 cm excluding the tail, which is about 10 cm long. It inhabits the North American forests from Canada to Mexico. A fold of skin stretched along each flank between the front and hind limbs helps the animal make gliding jumps up to 50 metres long. It is exclusively nocturnal.
Glaucomys volans
Sciuridae

African Ground Squirrel (130) inhabits the arid savanna and steppes of equatorial Africa. Its coat consists only of scarce bristles without underfur; there is a white stripe on its sandy flanks and its tail is mixed black and white. The related species, *X. rutilus*, lacks the stripe. A terrestrial animal living in small colonies, it digs out burrows in the manner of sousliks or prairie dogs.
Xerus erythropus
Sciuridae

European Beaver (131, 132) is the largest rodent of the Northern Hemisphere. It measures about 130 cm with the tail and weighs 30 kg, exceptionally 40 kg. The Beaver is perfectly adapted to aquatic life: it has waterproof fur with long hair and thick underfur, the webbed hind feet have enlarged paws and the horizontally flattened, paddle-like tail is scaly, used as a rudder in swimming. When diving, it can close all body orifices, including a sort of cloaca, that is a cavity into which intestinal, genital and urinary tracts open. It can stay underwater for up to 15 minutes. It has strong, gnawing teeth, red-brown in front; by gnawing it is able gradually to fell trees of up to 70 cm in diameter. Of all rodents, it has the largest cerebral cortex, and with the exception of the Norwegian Rat, it surpasses all of its allies in mental capacity. Beavers used to be plentiful in Europe and northern Asia; nowadays, they are confined to isolated sites near the lower Rhone
Castor fiber
Castoridae

131

The Beaver 'felling' a tree

132

in France, to the confluence of the Mulda and the Elbe in Germany, to Scandinavia, northern Poland, north European Russia, to Siberia and to limited areas in Mongolia and China. The decimation of beavers was largely caused by men, hunting them for fur, meat and mainly for the so-called castor (castoreum), a thick substance obtained from glands of the groin, used as a multi-purpose medicine. At present, beavers are protected and are being introduced into new localities.

Shallow forest brooks and river banks with thick shrubbery are the Beavers' favourite haunts. They build elaborate burrows in banks, or more often the so-called lodges, cone-shaped heaps of twigs and mud, with a diameter of 3–6 metres, on the water away from the banks. One or more chambers are located in the middle of these constructions, above water level. Beavers build the most elaborate dams of the same materials, sometimes several dozen or even hundred metres long: they impede the flow of water and raise the water level (132). Beavers live in permanent pairs or together with several generations of offspring, and form small family clans (up to ten members). Females bear one to five young, usually three, which can see immediately and are very independent. Diet consists of bark of young twigs and leaves, mostly of poplars and willows, and of various aquatic plants. Beavers do not hibernate but instead store up food in the water; they can later consume it under the ice sheet. The European and Canadian (*C. canadensis*) Beavers are the only living members of the beaver family.

133

Merriam's Kangaroo Rat
Dipodomys merriami
Heteromyidae

Merriam's Kangaroo Rat (133) represents an American family (Heteromyidae) of rodents which shows a typical convergent evolution with another, evolutionarily distant, non-American family — that of the Dipodidae, including jerboas. The two are almost identical in appearance. The Kangaroo Rat has the same elongated hind limbs with a highly reduced or lacking first toe; the

134

The Scaly-tailed Squirrel in a parachute jump

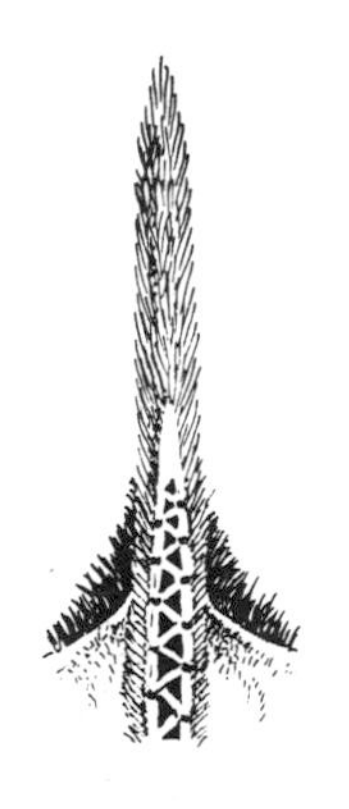

Underside of the tail of the Scaly-tailed Squirrel

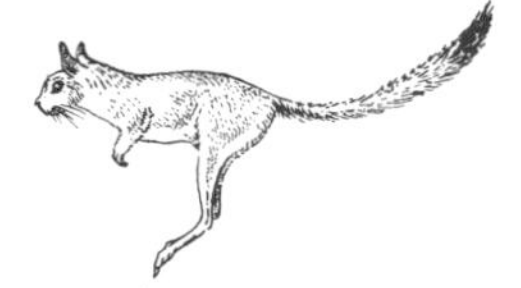

Jumping Spring Hare

135

paws are covered by fur, and even the skull is triangular, with large drum-like cavities as in jerboas. Kangaroo Rats inhabit desert and semi-desert places with scarce vegetation. They are nocturnal and feed on seeds, fruits and leaves of desert plants, and partly on insects. They also store up food in burrows. They never drink: water contained in the food is sufficient for them. Merriam's Kangaroo Rat is resident in Nevada, southern California, Arizona, New Mexico, Texas and Mexico.

136

137

Pel's Scaly-tailed Squirrel (134) is one of the 12 species of scaly-tailed squirrels, African
Anomalurus peli
Anomaluridae

arboreal rodents. It took its name from the horny scales covering the underside of the base of the tail. A wide fold of skin along each flank, from neck to limbs and tail, enables them to glide from tree to tree. They are diurnal and vegetarian. Pel's Scaly-tailed Squirrel is widespread in tropical forests from Sierra Leone to Ghana.

Cape Spring Hare (135) is one of the two species of spring hares. It measures about 40 cm in
Pedetes capensis
Pedetidae

length, with a tail the same length, and weighs up to 4 kg. It resembles a small kangaroo. The elongated hind legs enable it to jump 3 to 6 metres, exceptionally as much as 10 metres. Spring hares live in colonies, in underground burrows. They are nocturnal and vegetarian and live in southern and eastern Africa.

The rodent suborder Myomorpha, with some thousand classified species, is the most diverse group of rodents. They are mostly small animals with an advanced anatomy of the muscles of mastication, and usually with only three molars in each half of the lower jaw.

Deer Mouse (136) is common over almost all of North America, except Alaska and northernmost
Peromyscus maniculatus
Cricetidae

Canada; it reaches southwards as far as Mexico. It weighs 15–50 g. The 55 or so species of the genus *Peromyscus* are among the most abundant small mammals of the North American continent. They are ecologically equivalent to the Old World mice which are absent from the New World.

Burrow of the Common Hamster

138

Common Hamster (137) is the best known member of the hamster family, mainly for its proverbial tendency to store up quantities of food. It measures 20–30 cm, and weighs 400–700 grams. Its four-coloured coat is its typical feature; some individuals can, however, be entirely black or light-coloured. The mouth cavity contains large cheek pouches, in which food is carried. Originating in the steppes of eastern Europe and western Asia, the Common Hamster penetrated west with the expansion of agriculture, and settled in central and western Europe as far as Belgium and Luxembourg. It occurs only in lowlands and hills up to 500–600 metres of altitude, mostly in fields, pastures and shrubby forest steppes. Hamsters build intricate underground burrows with a nesting chamber and storage

Cricetus cricetus
Cricetidae

139

140

rooms. Following 19—20 days of gestation, females bear four to 12 young, often as many as three times a year. Hamsters live solitarily, the females staying with their offspring for only a short time. In favourable conditions, hamsters can overmultiply and become serious pests.

Florida Wood Rat (138), larger than the Common Hamster, in its appearance resembles a rat,
Neotoma floridana
Cricetidae
but has a characteristic bushy tail. It is a good climber, and builds spherical nests of twigs in trees. It occurs in the south-eastern USA.

Golden Hamster (139) has been domesticated either for laboratory research, or, more frequently,
Mesocricetus auratus
Cricetidae
as a pet. All golden hamsters kept in captivity are descendants of one family, caught in 1930 near the city of Aleppo in Syria. The Golden Hamster lives wild in Dobruja, Caucasia, western Asia and Asia Minor.

Roborowsky's Dwarf Hamster (140) is a little known representative of small hamsters of the
Phodopus roborovskii
Cricetidae
genus *Phodopus*, distributed in four species in western Siberia, the south Baikal region, Mongolia and China. It has a short tail and furry soles to the hind feet. It does not hibernate in winter. The species illustrated is the smallest: it measures about 9 cm. It was discovered at the beginning of this century. Its favourite habitats are the sandy deserts and semi-deserts of central China, Mongolia and in the Tuvinian uplands in Russia.

Crested Hamster (141) measures 25—36 cm and weighs 600—900 grams. It has a bushy coat
Lophiomys imhausi
Cricetidae
with a mane along the back. It is active at night, climbs well but slowly, and can move on tree trunks head downwards. It lives on leaves, shoots, and fruits. Its haunts are the thick, mountainous forests of eastern Africa, from Sudan to Somalia, Kenya, Uganda, Tanzania and Ethiopia.

Water Vole (142) is one of the relatively large voles (it weighs 100—300 grams). It usually lives
Arvicola terrestris
Microtidae
in the vicinity of water, but occurs also in fields, meadows and gardens. Although it is not especially equipped for aquatic life, it is a good swimmer

141

and diver. It digs out a system of burrows with a nesting chamber in river banks, and spends most of its time there. It also builds spherical nests in tufts of sedge in stagnant waters, using aquatic vegetation. It is chiefly nocturnal and feeds on various plants and invertebrates; it is known to attack small fish. In some places, it causes considerable damage by burrowing and gnawing at roots of young fruit and forest trees. Like all voles, it is extremely prolific, averaging three or four litters yearly with four to eight young. The Water Vole is widely distributed over Europe, and in central Asia to Mongolia and Baikal.

Common Vole (143)
Microtus arvalis
Microtidae

is a typical member of the most numerous genus of voles, widespread in the forest and steppe zone of North America and Eurasia. In Europe it is regarded as a dangerous harvest pest. Although native to Oriental forests and steppes, it adapted well to life in agricultural regions. Its favourite haunts are fields, meadows and roadside verges, but it also

142

143

penetrates into mountains, forests and marshlands. The diet consists mainly of green parts of plants. The Common Vole is gregarious. It digs elaborate systems of burrows, not very far under the surface, with exits connected by tracks. It is very prolific, being able to produce three to seven litters of one to 13 young. It breeds even in winter if conditions are favourable. Its reproductive potential is sometimes increased by a reduced time to sexual maturity, the birth of a greater proportion of females, and the communal rearing of offspring of several mothers. This triggers off periodic overpopulation, known as 'mouse years'.

Muskrat (144) is a North American rodent, introduced in 1905 to Bohemia, from where it spread quickly throughout all central Europe. Later it was imported to western Europe, Scandinavia and Russia. Nowadays it is distributed in a major part of Europe and in vast areas of Siberia, Transcaucasia and Kazakhstan, penetrating as far east as Mongolia and China. In its native continent, it ranges from Alaska to the Atlantic coast, and to California and Mexico. The Muskrat cannot survive away from banks and shrubberies near rivers and stagnant waters. Swimming and diving is facilitated by the thick, waterproof fur, broad paws of the hind feet lined with coarse bristles, and mostly by the long, laterally flattened tail, which is almost hairless. Burrows and nests are built in banks; the so-called lodges are constructed of roots and stems of aquatic plants in shallow water with vegetation. The females produce three or four litters a year, averaging five to seven young. Muskrats supplement their vegetarian diet with mussels, crayfish and exceptionally fish, mostly dead ones. This is the largest species of vole, measuring 44–68 cm with a tail 18–28 cm long and weighing 0.7–2.4 kg.

Ondatra zibethicus
Microtidae

Bank Vole (145) has a characteristic, rust-coloured back, and a relatively long tail. Its total length is 13–15 cm. It inhabits forests with undergrowth, clearings and dense shrubbery. It is widespread almost everywhere in Europe except the northernmost parts, and in the mountains of Transcaucasia, central Asia and Asia Minor. Related species live in northern Europe, Siberia and

Clethrionomys glareolus
Microtidae

144

North America. It forms family groups and breeds three or four times a year, each time producing three to five young. It occasionally overbreeds. It is active both night and day and is mostly terrestrial, but it can also climb trees. Nests of grass, moss and leaves are built in underground burrows, under fallen trees, under stones, and in roots. It eats mostly green parts of plants, seeds, roots, bark (in winter) and insects.

Southern Jird (146) occurs from south-eastern Europe over central Asia to Mongolia and northern China. It lives predominantly in sandy deserts and semi-deserts, digs out deep burrows, and feeds on seeds, rhizomes, bulbs and green parts of plants. It is one of the plague-carriers.
Meriones meridianus
Microtidae

Lesser Mole Rat (147) belongs to the family of Spalacidae, including five species from south-eastern Europe, western Asia, and north-western Africa. It is curiously adapted to underground life: it has a cylindrical body covered with very short, thick and glossy fur, covered-over eyes, completely reduced ears and a flat head with formidable protruding incisors. Except for storerooms and winter shelters, its burrows are built near the surface, and the earth is piled above ground in large molehills. The Mole Rat does not use its feet for digging, unlike most underground mammals: it uses its strong incisors. It is found in fields, prairies, steppes and screes from lowlands to high mountains. It feeds on underground parts of plants and hardly ever goes above ground. It is widespread in the Balkan Peninsula, the Ukraine and Asia Minor.
Spalax leucodon
Spalacidae

The Bank Vole gnawing at a hazel nut

145

Black Rat (148) belongs to the family of Old World rats and mice, the most numerous rodent
Rattus rattus
Muridae

family in terms of species: about 400 of them have been classified and new ones are still being described. Evolutionarily, this family ranks among those rodents of rather recent derivation; they appeared in the Miocene and still are developing. Rats are probably native to tropical south-eastern Asia; their dependence on Man and his settlements caused them to spread throughout the world. This rapid spread is evident from findings of remains from the inter-glacial eras in western Europe and by other fossils, several thousand years old, from the Mediterranean. The Black Rat reached America probably with the first ships. The ever-increasing naval transport helped it to spread all over the world, travelling in the cargo ships it infested. Black Rats are warmth-loving animals, in-

146

147

habiting mainly wooden buildings, and preferring upper floors and attics. In warm regions, for example in the Mediterranean, they live in the open; in central Europe they are bound exclusively to human settlements. In the Middle Ages, the Black Rat became the plague of Europe and was the main cause of repeated epidemics of bubonic plague and other infectious diseases. At present, it is relatively scarce in Europe, being replaced there by its larger relative, the Norwegian Rat (*Rattus norvegicus*). Still, its decrease in numbers is more probably a result of the elimination of wooden buildings than of competition between the two species. The Black Rat differs from the Norwegian Rat by its smaller size (13—24 cm), weight 140—250 grams), comparatively longer tail, longer ears and more pointed muzzle. It is not as omnivorous as the Norwegian Rat, preferring plant food.

148

149

House Mouse (149) is Man's unwelcome companion in villages and towns. Of all rodents, it is
Mus musculus
Muridae

the one most responsible for man's negative attitude towards them. It was originally restricted to the Asian and east European steppes, feeding there on seeds of steppe grasses. With the expansion of agriculture, it spread to almost all of Asia and Europe, as well as to other continents. In colder regions it is confined only to human settlements, whereas in warmer parts it lives in the open as well, mostly in fields. It is omnivorous and causes great damage by devouring and spoiling stored grains and food. It also leaves an unpleasant smell and spreads certain contagious diseases. It breeds often throughout the year, and produces yearly five to seven litters of four to eight young. Its total length is only 150 to 180 mm and it weighs 15–28 grams.

Yellow-necked Field Mouse (150) is a common representative of the long-tailed, wild mice of
Apodemus flavicollis
Muridae

the genus *Apodemus*, living mostly in forests, woods and fields. The Yellow-necked Field Mouse is the largest species: it measures 90 to 130 mm, with a tail of the same length or longer, and weighs 15–40 grams. On the flanks the colour of the upper parts is in sharp contrast to the white abdomen. It is an excellent runner, jumper and climber. It inhabits both deciduous and coniferous forests from plains to high mountains; in winter, it sometimes moves to buildings. Shelters are made in underground nests, hollows, etc. Food consists mostly of seeds of forest trees and of insects. It is a nocturnal animal. The female has three to eight young two to four times a year.

Egyptian Spiny Mouse (151) has tough, spiny hairs on its back, and a rather short tail. It weighs
Acomys cahirinus
Muridae

30–70 grams. Among rodents of the same size, it has the longest gestation period: 36–38 days. The young are born hairy; they can see, and move out of the nest on the third day. Females assist each other when the young are born, and rear them together. The Egyptian Spiny Mouse is resident in stony areas, in steppes and deserts of western Asia and northern Africa.

Striped Grass Mouse (152) inhabits equatorial Africa, from Sierra Leone to Tanzania. It measures 10–14 cm, the tail being approximately the same length. It is active in the day and early evening, and lives on seeds, green parts of plants, and insects.
Lemniscomys striatus
Muridae

Fat Dormouse (153) is the best known representative of the dormouse family, distributed in 25 species in forest and forest-and-steppe regions of Europe, Asia and Africa. The Fat Dormouse is the largest species, measuring 130–180 mm without the tail, and weighing 70–120 grams. It is common in the light and warm deciduous forests from Spain over southern and central Europe to the Caucasus, Asia Minor and northern Iran. It has been successfully introduced into the British Isles. The Fat Dormouse has become adapted to life in parks, orchards and gardens. It eats fruits, buds, leaves and insects; occasionally even young birds and birds' eggs. It often finds shelter in cottages or birds' nest boxes, or it builds spherical nests of leaves and twigs in trees and bushes. It is a strictly nocturnal species. The winter season from October to April or May is spent in hibernation: the dormouse rolls into a ball in a sheltered place.
Glis glis
Gliridae

Garden Dormouse (154) is the most colourful of all dormice. It is common in the warm Mediterranean; in central Europe it has a scattered distribution, reaching north as far as the Baltic region and central Finland, and eastwards as far as the Urals. More than other dormice, it moves on the ground, finding shelter in rocks, piles of stones, etc. Its diet consists mostly of insects.
Eliomys quercinus
Gliridae

150

151

Common Dormouse (155) reaches a mere 75—86 mm. It is widespread over almost all of Europe, reaching as far east as the Volga River and Asia Minor. It is the only dormouse indigenous to the British Isles. It prefers mixed forests, but lives also in various wooded areas from lowlands up to the dwarf pine belt of the mountains. It consumes chiefly plant food and insects. It builds spherical nests of leaves and moss in bushes, near the ground, or finds shelter in hollows. It is nocturnal and hibernates like all dormice.

Muscardinus avellanarius
Gliridae

Siberian Jerboa (156) represents the family Dipodidae, including some 23 rodent species, adapted to life in steppes and deserts of eastern Europe, Asia and Africa. They move by long leaps on the hind legs. Active at night, they hide in underground burrows during the day. Food consists of desert plants and insects. The species illustrated occurs from the eastern coast of the Caspian Sea and Ural regions to the southern Trans-Baïkal area, Mongolia and northern China.

Allactaga sibirica
Dipodidae

152

153

Northern Birch Mouse (157) belongs to the family Zapodidae, existing in several species in northern and eastern Europe, central and eastern Asia and North America. The Northern Birch Mouse is found in northern Europe and as a relict in certain mountains of central Europe.

Sicista betulina
Zapodidae

White-tailed Porcupine (158) is a member of the peculiar porcupine family, including some 15 species of large and medium-sized rodents, confined to Africa and Asia; only one species reaches southern Italy and Sicily. The White-

Hystrix leucura
Hystricidae

154

155

tailed Porcupine measures 75 cm and can weigh up to 15 kg. The back and sides are covered with long, thick, black-and-white quills, only lightly attached in the skin so that the animal can not only erect them, but also, when danger threatens, will rattle them by vigorous contraction of muscles of the back. The short tail has special, cup-shaped quills; when shaken they produce a rattling sound. The back of the neck bears a mane of spiny bristles; the abdomen, muzzle and limbs are covered with sparse, bristly hair. The White-tailed Porcupine inhabits arid, rocky places in Asia Minor, western and central Asia, India and Ceylon.

Naked Mole Rat (159)
Heterocephalus glaber
Bathyergidae

has become completely hairless during its evolution: the body is covered by a pink, wrinkled skin with sparse, individual hairs. Its body length is 8–9 cm, tail length 3–4 cm and the weight 30–45 grams. It lives underground and almost never leaves the burrows, which have funnel-shaped exits. Its diet consists of underground parts of plants and insects. It occurs in Ethiopia, Kenya and Somalia.

North American Porcupine (160)
Erethizon dorsatum
Erethizontidae

lives in forests of North America, from Alaska and Canada to Mexico. It is an arboreal species, and climbs and swims well. It is a solitary, nocturnal animal, feeding on bark, leaves, berries and seeds. Its usual weight is 4–7 kg, exceptionally as much as 18 kg.

For a long time, South American rodents have been developing independently of rodents from other continents, and they have formed a group, heterogeneous in terms of appearance, included in the suborder of cavy-like rodents (Caviomorpha). Their young are born well-developed, hairy, with open eyes and are independent.

156

Jumping jerboa

157

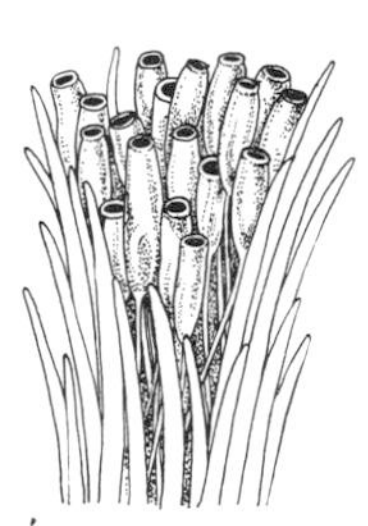

Cup-shaped quills found on the tails of some porcupines

158

Capybara (161) is the only species of its family, and the largest existing rodent. It measures more than 1 metre, reaches a height of 50 cm and has a weight of about 50 kg. Its hair is long but sparse. The Capybara has four toes on the forelimbs; the three on the hind limbs are connected by short webs. It lives in the dense, swampy primary forests of equatorial South America, eastwards from the Andes. It swims and dives well, and feeds on the green parts of plants.

Hydrochoerus hydrochaeris
Hydrochoeridae

159

160

Mara (162) measures about 70 cm and weighs 9—16 kg. It has long legs with hoof-like claws, and long ears. It inhabits open, arid and grassy places and is a fast runner and good jumper. It is a diurnal species and lives in small groups. Its diet comprises plant food. The Mara is restricted to southern Argentina.

Dolichotis patagona
Caviidae

161

162

Viscacha (163) is the largest species of the family of viscachas and chinchillas; it weighs up to 7 kg. It lives in colonies of 15 to 30 individuals, and builds complex systems of underground burrows. It is nocturnal, a good runner and jumper. It is noted for bringing various objects, particularly shiny ones, to its burrows. The Viscacha is found in the Argentinian pampas.

Lagostomus maximus
Chinchillidae

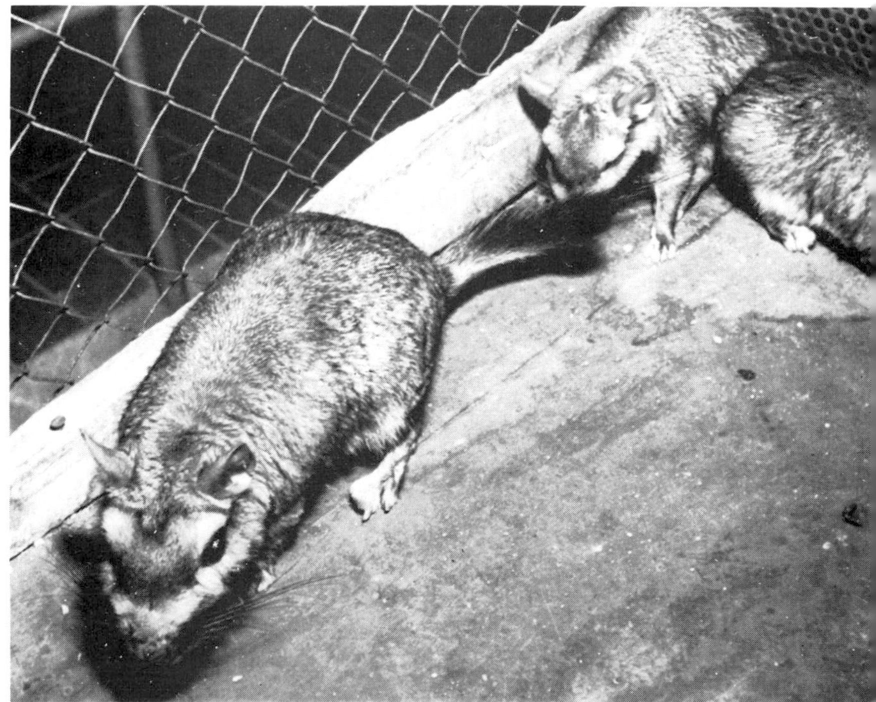

163

Chinchilla (164) is endowed with a highly valuable, thick, soft and silvery pelt, for which this formerly abundant animal had been almost exterminated. The last specimens live nowadays in Bolivia and Chile, high up in the mountains at 3,000–6,000 metres; they are strictly protected. It was from there, in the 1920s, that chinchillas were brought for farming, mostly to North America and Europe. Fur of the artificially bred chinchillas does not, however, reach top quality. Attempts at re-introduction of chinchillas into new localities in the Andes, or the Pamirs, have so far been unsuccessful. The Chinchilla measures 22–40 overall cm; the tail is bushy and 15–18 cm long. It has large eyes and long ears, and weighs about 0.4–0.5 kg. It lives in colonies, in rocky places with scarce vegetation. Females bear one to six young, one to three times a year.

Chinchilla laniger
Chinchillidae

Nutria (165) is adapted to semi-aquatic life; it seeks the banks of slow-flowing rivers and lakes, where it digs burrows, or builds reed shelters in shallow water. It is an excellent swimmer and diver. It weighs 7–9 kg, 14 kg at most. It has powerful, orange-coloured incisors, and its toes are connected by a short web. Its range covers South America from Paraguay and southern Brazil to the Magellan Strait; it has been exterminated in many places. In recent years, it has been introduced into various regions, chiefly in North America and the former USSR. It is kept on a large scale on farms, as it produces very valuable fur.

Myocastor coypus
Myocastoridae

164

165

Degu (166) represents the little known family of Octodontidae. Its total length is about 24–27 cm.

Octodon degu
Octodontidae

It lives in Chile, where it is quite abundant in places. Its way of life is mostly terrestrial.

166

167

Chapter 8 CARNIVOROUS MAMMALS OF THE LAND

The vital task of maintaining the necessary balance in Nature falls to the members of the order Carnivora, which comprises a group of nine families of mostly flesh-eating, less frequently omnivorous, animals. Their main function in Nature is to keep in check the numbers of herbivores, and also those of some other mammals. They hunt for food, and their body structure is perfectly adapted to this way of life. The brain and senses, particularly smell, sight and hearing, are usually highly developed. Carnivorous mammals can run fast and, except bears, most have long tails. They walk mostly on the toes alone (they are digitigrade), only exceptionally their whole soles touch the ground. Limbs have four to five toes with strong claws. They have conspicuous canine teeth, long and sharp, for catching and killing the prey. Many carnivores have the last upper premolar and the first molar of the lower jaw transformed into the so-called carnassial teeth which have a flattened, razor-like crown and serve for tearing muscle tissue. (In omnivorous carnivores, for example in bears, true carnassial teeth are not developed.) Their prey consists of various mammals, more rarely of birds and fish: it is not always caught easily, and carnivores guard it carefully. Unless they consume it immediately, they drag the food into a shelter, bury it in the ground, or even drag it up a tree (167). Carnivores usually produce several young, born underdeveloped and requiring the care of the mother. With the exception of the dog family, living and hunting in packs, carnivores are mostly solitary, except during the breeding season. They are chiefly nocturnal, or active at dusk; they are terrestrial or partly arboreal, or even aquatic. Australia is the only territory where original carnivores were lacking. Nowadays, carnivores cover all geographical zones, and their numbers depend on the distribution of their prey.

Common Weasel (168)
Mustela nivalis
Mustelidae

is a member of the weasel family, smallish carnivores with slender, elongated bodies and short legs. They walk on the toes or front part of the soles; the toes have short, sharp claws. Scent glands are situated under the tail: the secretion can be sprayed some distance. The scent is a sexual attribute, and it marks the limits of the animal's territory. The majority of weasels are carnivorous: the staple food consists of mice and voles; only a few exceptions are omnivorous. They provide valuable fur, for which they are often ruthlessly hunted. The Common Weasel is

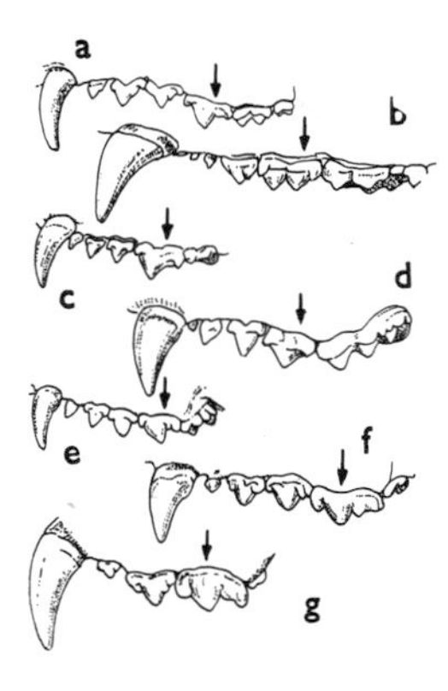

Teeth of various beasts of prey: a) wolf, b) bear, c) marten, d) badger, e) mongoose, f) hyaena, g) lion. Arrows indicate carnassial teeth

168

169

about 20 cm long; it is distributed over Europe, Asia, North America and northern Africa. It lives in prairies and fields, at forest edges, near human settlements, and in the north even in open tundras. It avoids thick forests and damp ground.

Ermine or **Stoat** (169, 170) differs from the Common Weasel in having a black tip to its tail, and being slightly larger. In winter, the Ermine's coat becomes snow-white, only the tip of the tail remaining black (170). Its fur, also called ermine, made the Stoat a much-hunted animal, particularly in the past. It is found in Europe, northern Asia, Japan, and North America. It feeds mostly on mice, lemmings, voles and other small rodents; exceptionally on birds or larger mammals, e.g. hares. The gestation lasts 8 weeks if mating takes place in early spring. In May or June five to seven young are born. The second mating season is in June and July; in this case the development of the embryo is delayed, and offspring are again dropped in May.

Mustela erminea
Mustelidae

170

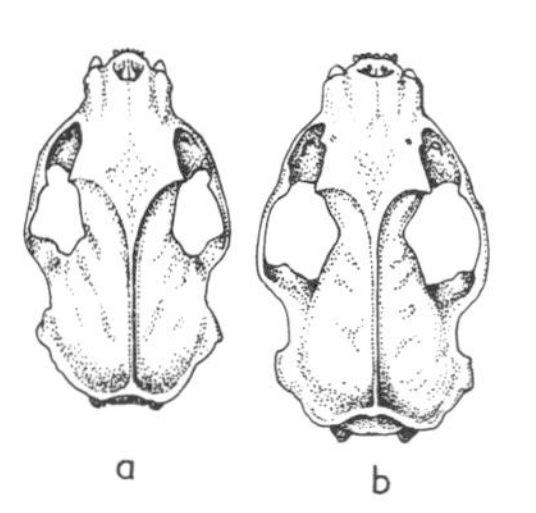

Skulls of the Common Polecat (a) and Eversman's Polecat (b) demonstrate a pronounced difference in the structure of the postorbital area

171

Common Polecat (171) attains 40–45 cm in length, the tail is 10–15 cm long, and its weight may surpass 500 grams. It occurs almost all over Europe except Ireland, part of England, northern Scandinavia and southernmost Europe. It inhabits mostly mixed and deciduous woods, shrubberies and groves, often close to human dwellings. It does not even avoid the confines of large cities. It eats chiefly small rodents, but also birds up to the size of a duck or pheasant, birds' eggs, etc. Frogs and fish seem to be polecat's favourite food: it traps them directly in shallow water. Foraging trips take place mostly at night.
Putorius putorius
Mustelidae

Eversmann's or **Steppe Polecat** (172) is closely related to the preceding species, differing from it by the tawny colouration and by its somewhat larger size: its distribution also is different. It is resident in central and south-eastern Europe, southern parts of the former USSR, Mongolia, northern China, and the Far East. It prefers open steppes or inhabited areas with fields and dry meadows. In central Asia, it is found at altitudes of 2,500 metres above sea level. Its diet consists of steppe rodents, occasionally also steppe hares and birds. Like the Common Polecat, it makes stores of food in its dens. Both species mate mostly in spring; gestation lasts about 40 days.
Putorius eversmanii
Mustelidae

American Pine Marten (173) represents the genus of true martens in the north-east and far north of North America. It reaches a length of about 65–70 cm, including the 15–17 cm long tail. It feeds on small rodents and birds. Although it is called the 'American sable', the quality of the pelt cannot compare with that of the true North Asiatic Sable (*Martes zibellina*), reputed for its valuable fur.
Martes americana
Mustelidae

172

Pine Marten (174), with the American Pine Marten and the Sable, is a typical inhabitant of dense forests. It occurs almost all over Europe, and in western Siberia. The Pine Marten has a fawn or deep-brown coat, always with a conspicuous, yellowish or orange yellow spot, narrowing to one lower point, on the neck and chest. The body measures more than 80 cm including 25 cm of tail. The Pine Marten lives on various mammals and birds, up to the size of a hare, and on forest fruits.

Martes martes
Mustelidae

Beech Marten (175) is slightly smaller than the Pine Marten. Colouration is similar, but the spot on neck and chest is always white and has two tips reaching the inside of the upper part of the forelimbs. It is less restricted to forests, preferring

Martes foina
Mustelidae

173

174

rather open and rocky places; it can be found near human settlements, even in city parks. It occurs from central and southern Europe and Asia to northern China, sometimes up to 4,000 m of altitude. Its diet resembles that of the previous species, but the Beech Marten also attacks poultry, and sometimes will even kill all the hens in a hen-house without devouring them. As in the Ermine, latent pregnancy is a common feature in martens and lasts, according to species, from 250 to 290 days. Average litters vary between two and eight young; three to four is usual.

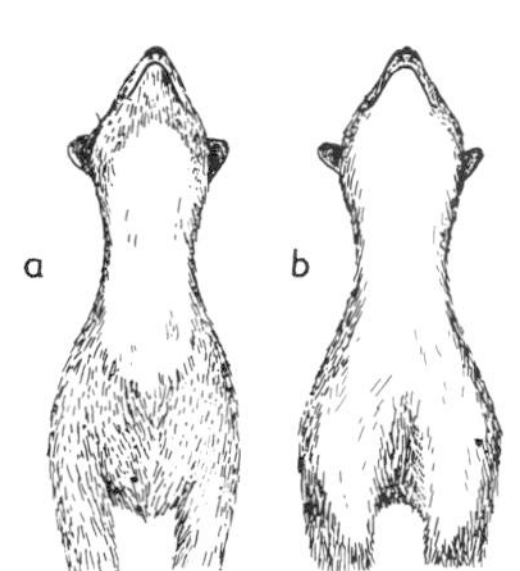

Different shapes of the pale spot on throat and chest of the Pine Marten (a) and the Beech Marten (b)

175

176

Kharza Marten (176) measures up to 130 cm, including the 45 cm long tail, and the weight is about 6 kg. It inhabits the forests of southern, south-eastern and north-eastern Asia from India, Greater Sundas and Borneo (Kalimantan), to the Ussuri River. It feeds not only on small mammals and birds, but sometimes kills even the young of smaller hoofed mammals.

Lamprogale flavigula
Mustelidae

Grison (177), an inhabitant of Central and South America, resembles a polecat in its size and other features. It is most common at forest margins and in plantations. It does not avoid buildings, and sometimes can cause serious damage to poultry farms.

Galicitis vittata
Mustelidae

Wolverine (178) is one of the largest representatives of the weasel family. The body measures up to 90 cm, the tail 20 cm, shoulder height is 45 cm and the weight more than 30 kg. Colouration is greyish to reddish brown, with a dark saddle-like patch on the back; underparts and limbs are also dark. It is resident in northern Europe, Asia and northern parts of North America, inhabit-

Gulo gulo
Mustelidae

177

178

ing northern forests, taigas and tundras. It preys on the young and small specimens of reindeer, moose, deer and other hoofed animals, and on rodents and birds; on occasion, it may even kill foxes and otters. It also eats various fruits, but to a limited extent. Gestation lasts 8–9 months; there are one to four young in a litter.

Honey Badger (179) is about 1 m long, including the tail, which measures 18–20 cm. Its weight
Mellivora capensis
Mustelidae

does not usually surpass 14 kg. The upper side of the body is ash-grey, underparts are black-brown. The Honey Badger lives in various places in Africa, western and central Asia and India, mostly in rocky and steppe-like locations. Its diet is composed of small rodents and birds, also of snakes, frogs and many invertebrate species. The Honey Badger is a good climber. Its favourite delicacy, honey of wild bees, is skilfully dug out of the bees' nests. It is noted for its co-operation with the woodpecker-like bird, the Honey Guide (*Indicator indicator*) which leads it to nests of honey bees. The Honey Badger breaks up the nest and devours the honey, while the Honey Guide picks up the larvae.

179

180

Old World Badger (180) is well-known for its hibernation which, however, differs from the true hibernation of bats or some rodents. In badgers, the body temperature does not decrease, and the animal wakes up from its sleep several times during the winter. Adult Old World badgers grow up to the size of honey badgers, but in autumn, before settling to hibernate, they may weigh more than 25 kg. Except Sardinia and Scandinavia, the Old World Badger inhabits all Europe, central Asia and northern parts of Asia to the Amur delta. It is also found in Japan. Badgers show a preference for

Meles meles
Mustelidae

181

Menacing attitude of a skunk, before going into 'chemical warfare'

182

dense woods. They feed on invertebrates, small rodents, birds and their eggs, and largely also on plant food. The phenomenon of latent pregnancy, lasting approximately 9 months, also occurs in badgers.

American Badger (181), also called 'tejone', is conspicuous by the apparently flattened back:
Taxidea taxus
Mustelidae

this is caused by the length of the fur and direction of its growth. The American Badger inhabits most of the North American continent, from Mexico to approximately 60°N. It does not differ greatly from its Old World counterpart in way of life and diet.

Common or **Striped Skunk** (182) is roughly the size of a polecat; it is widespread over North
Memphitis memphitis
Mustelidae

America, reaching as far north as the Hudson's Bay. It seeks shrubby and wooded places, particularly near water. It feeds on small vertebrates and various invertebrates. The Common Skunk is notorious for its smell; it sprays a secretion of the scent gland from vents under the tail. This purely defensive chemical weapon is carefully avoided by all animals, including men, at all costs: they all prefer to leave the skunk alone.

Common Otter (183) is bound to water by its way of life. Its body has a hydrodynamic shape,
Lutra lutra
Mustelidae

and the feet are webbed. It builds burrows with underwater entrances in river banks. With its tail of 40—50 cm the Common Otter reaches a length of 120—145 cm, and a weight of some 15 kg. It lives in Europe, northern Africa, a large part of Asia, and in North America. Its diet consists mostly of fish, skilfully trapped in water, but otters also catch frogs, less frequently even small mammals and birds. Gestation lasts about 9 weeks or about 9½ months (latent pregnancy).

183

Sea Otter (184) is the largest member of the weasel family. Adults may measure up to 180 cm including the tail, which is 35–40 cm long. Old males weigh as much as 42 kg. The Sea Otter became almost extinct because of its precious fur; it is no longer endangered thanks to strict protection in the past decades. It is an excellent swimmer and diver: the hind limbs resemble, in their structure, those of pinnipeds. It lives in large or small groups on the Pacific coasts of north-eastern Asia, chiefly on the Commander Islands, and off the North American coasts, from the Aleutian Islands and Alaska to California. It feeds mostly on fish and sea urchins, occasionally on crabs and various molluscs. Gestation lasts 8–9 months: unlike other

Enhydra lutris
Mustelidae

184

185

weasels, the female produces a single young (exceptionally two). Colouration of the Sea Otter varies with age: the original brown to blackish-brown colour in juveniles gains, in older specimens, a silvery to silvery-orange tone on the head and chest and part of the back.

Ringtail or **Cacomistle**
Bassariscus astutus
Procyonidae

(185) represents the family Procyonidae, including typical plantigrades. The family is exclusively omnivorous. They are usually not larger than a fox, always with a long tail (that of the Ringtail is 40 cm, its body 50–55 cm). Their skull structure bears numerous primitive features by which they differ from all other carnivores. The Ringtail lives in Mexico and the south-western United States. It is a nocturnal species and feeds on small animals and plants.

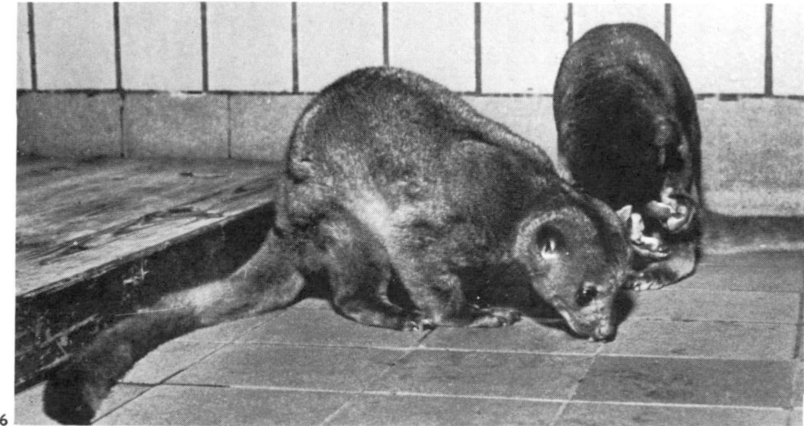

186

187

Kinkajou (186) is about the size of or slightly larger than the Ringtail. Its colouration is grey-ochre above and lighter below. The tail is prehensile. The Kinkajou inhabits the tropical primary forests of Central and South America, from southern Mexico to the Brazilian Mato Grosso. It eats mostly tropical fruits, birds' eggs and young, insects and the honey of wild bees. It is nocturnal, keeps mostly to trees and is an excellent climber. Pregnancy takes 16–17 weeks; only two young are born.

Potos flavus
Procyonidae

188

189

Ring-tailed Raccoon
Procyon lotor
Procyonidae

(187) is the best-known member of the family Procyonidae. In many languages, its name is derived from the habit of washing various objects in water. It is about 100–105 cm long, 35–40 cm of this being the tail. The fur is yellowish-grey with individual black hairs. A conspicuous black band crosses the eyes, creating the typical raccoon mask. The Ring-tailed Raccoon lives in Central and North America, including the southern parts of Canada. It settles in thickets on the banks of rivers and water reservoirs. It is mostly carnivorous. Following 63–65 days of gestation, the female bears one to seven young. In winter, raccoons go through a period of false hibernation.

190

191

Crab-eating Raccoon (188) is a South American relative of the Ring-tailed Raccoon. It is slightly larger, of lighter colour, and its fur is shorter and smoother. Its way of life is almost identical. It ranges from Panama and Costa Rica to northern Argentina.

Procyon cancrivorus
Procyonidae

192

193

Red Coatimundi (189) reaches some 120 cm in length, about a half of this being the tail. It is widespread over South America, in the mountain areas close to the western coast reaching up to 3,000 metres of altitude. It lives both in wooded regions and in treeless country, even on the margins of pampas.

Nasua nasua
Procyonidae

White-lipped Coatimundi (190) is about the same size as the preceding species. In colouration, grey prevails over the reddish tones, and the muzzle and area around the eyes are boldly patterned. The White-lipped Coati is widespread in Central America and in the southernmost parts of south-west North America. Both species live in groups; they feed on insects and other invertebrates, small mammals and birds, birds' eggs, and various fruits and soft parts of plants. Gestation lasts about 75 days; litters average two to seven young.

Nasua narica
Procyonidae

Lesser Panda (191) measures up to 110 cm including the tail, which is about 50 cm long. It is distributed in Nepal, Sikkim, Bhutan, Assam, and in south Chinese Yunnan and Sechouan provinces, where it can be found even at altitudes over 4,000 metres. It leads a secluded life in forests and bushes, where it feeds on birds, their eggs, insects and tender parts of plants, preferring young bamboo shoots. Gestation lasts approximately 130 days; the female bears one to four young. With the following species, the Lesser Panda has recently been classified into a special family of pandas (Ailuridae), comprising only the two species.

Ailurus fulgens
Ailuridae

Giant Panda (192) — despite its rarity — is today one of the best known and most popular mammals. It was even selected for the emblem of the World Wildlife Fund, as a symbol for the protection of wildlife. The Giant Panda eats bamboo shoots almost exclusively; other food, for example insects and eggs, is taken in minimum quantities. Accordingly, it is sometimes called

Ailuropoda melanoleuca
Ailuridae

194

'bamboo bear'. Its length may exceed 155 cm, and the weight is 120 kg or more. A single young is produced. The Giant Panda occurs only in the inaccessible areas of the Sechouan Province in western China, and on the adjacent Chinese-Tibetan border. It may only very rarely be seen in captivity.

195

196

Brown Bear (193 – 195) *Ursus arctos* Ursidae

is a typical member of the bear family which are often very bulky animals and include the largest living carnivores. They are plantigrades, and the structure of their teeth and skeleton implies that they are not greatly adapted for hunting. Plants form a considerable part of their diet, but they are considered to be omnivorous. The Brown Bear originally in-

197

198

habited all of Europe, parts of the African Atlas Mountains, much of Asia north of Syria, Lebanon, Iran and the Himalayas, and vast territories of North America. Nowadays it has become extinct in many of these places. Widely dispersed over such an immense territory, the Brown Bear formed a series of subspecies differing in colour, size, and details of the skull. The best-known is the European Brown Bear (*Ursus arctos arctos*) (193), reaching a length of 2 metres and weighing up to 300 kg.

199

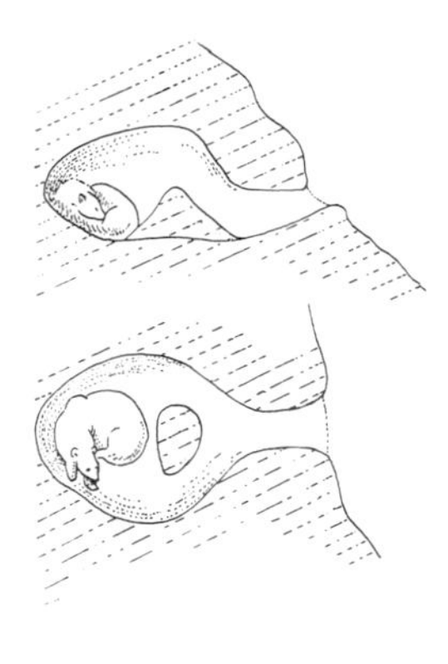

Winter den of the Polar Bear

200

The North American Grizzly (*Ursus arctos horribilis*) (194) has a greyish coat, is more than 2 metres in length and weighs almost 450 kg. A real giant, and in fact the largest living carnivore, is the Kodiak Bear (*Ursus arctos middendorffi*) (195), living on Kodiak Island and some other islands

201

202

off the southern Alaskan coast, and occurring even on the Alaskan Peninsula. It reaches a length of 280–290 cm, and may weigh up to 900 kg. The Brown Bear seeks dense forests with rocks, uprooted trees, etc. It can be found high in the mountains above the tree line, even at 4,000 metres altitude. In terms of plant food, it prefers buds, fruits and tender parts of plants; the meat component is derived from small mammals, birds, reptiles, molluscs and insects. It searches diligently for nests of wild bees: their honey is a delicacy for bears. Alaskan bears are noted

203

also for their skill in salmon hunting. In winter, bears fall into a state of torpor which is called false hibernation; the body temperature keeps at normal values. Bears wake up several times during this period and even leave the den. When hibernating, they live off stores of subcutaneous fat, accumulated mainly in the autumn. The young (mostly two, but up to six) are born during the winter lethargy, after a latent pregnancy of 8–9 months, and the female suckles them and warms them until spring, when they leave the winter shelter for good.

American Black Bear
Ursus americanus
Ursidae

(196) is substantially smaller than the Brown Bear. It has a glossy black coat, sometimes with predominant cinnamon or silvery tones. It is resident in North America, from central Mexico to Alaska and Labrador. In its habits, it is similar to the Brown Bear. Gestation lasts 6–8 months, the female produces one to four (usually two) young.

Asiatic Black Bear
Selenarctos thibetanus
Ursidae

(197) is distinguishable by its rounded ears and long, black hair with a single, V- or Y-shaped bright white spot on the chest. Old males reach a length of 160 cm and weight of 120 kg. The Asiatic Black Bear is widespread in southern and south-eastern Asia, in China, Japan, Korea, in the Ussuri and Amur regions, and on the islands of Hainan and Taiwan. It shows a preference for mountain forests. It is omnivorous: in some areas plants take precedence over flesh, elsewhere it is the reverse. Asiatic Black Bears climb trees well and sometimes even fabricate nests of sorts in tree tops. Gestation lasts 5–6 months; usually two young are born.

Sloth Bear
Melursus ursinus
Ursidae

(198) inhabits forests of India, and Sri Lanka. Its long, shaggy coat has a matt black colouration, or is black-brown with reddish overtones; there is a U-shaped white spot on the chest. The body is up to 180 cm long; weight is 140 kg at most. Besides plant fruits and shoots the Sloth Bear eats chiefly insects and larvae, honey of wild bees, birds' eggs and carrion. It is renowned

205

for its remarkable way of catching termites: the animal starts by digging out the termitary with its formidable, sickle-shaped claws, then blows away dust and debris, and sucks termites into the mouth like a huge vacuum cleaner — simply by sucking in the air. This method of hunting is facilitated by many adaptive features: the lips are long and naked and are able to form a tube; the nostrils can be completely closed; the palate is hollow, and the front pair of incisors is missing, so that the bear can comfortably suck through the space between its teeth. The sound emitted by bears 'blowing' at a dug-out termitary and sucking out its inhabitants is so strong that it can be heard at a distance of 200 metres. The Sloth Bear does not hibernate. Gestation lasts about 7 months; one or two, exceptionally three, cubs are born.

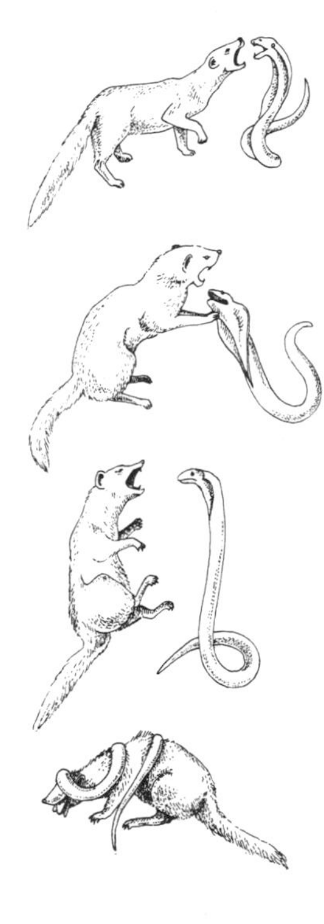

The Common Mongoose fighting with a venomous cobra

206

207

Malayan Sun Bear (199) is the smallest bear; it weighs scarcely 70 kg. Its fur is short, glossy black and, unlike other bears, very smooth. The muzzle is greyish or slightly orange-coloured. There is a white spot, usually shaped like a wide U, on the chest. It is a good climber and likes to build nests in trees. Its diet is similar to that of Sloth Bears. It lives in Burma, Indo-China, on the Malayan Peninsula, in southernmost China, Sumatra and Borneo (Kalimantan). The Malayan Sun Bear does not hibernate. After 95—100 days of gestation, the female gives birth to one or two young.

Helarctos malayanus
Ursidae

Spectacled Bear (200), also called **Ucumari,** is the only bear on the South American continent. It lives in the wooded mountainous areas of western Venezuela, in Colombia, Equador, Peru, and western Bolivia. Body length is up to 1.8 metres and weight about 140 kg. The coat is black or black-brown

Tremarctos ornatus
Ursidae

208

209

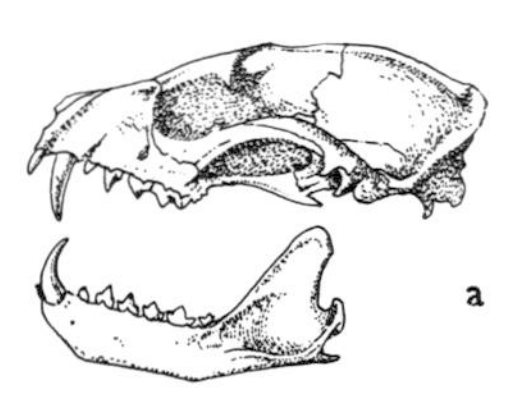

a

b

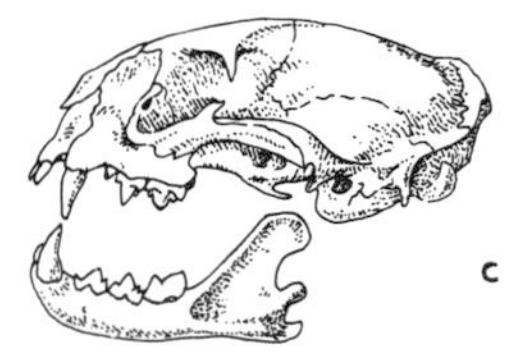

c

The skull of the Fossa (b) distinctly differs from skulls of typical civet carnivores, for example, the Palm Civet (a), and resembles more closely the skulls of felines, for example the Leopard Cat (c)

210

with conspicuous, light or white facial markings, descending to the throat and chest. The pattern creates a conspicuous spectacle-shaped mask around the eyes. The Spectacled Bear is a good climber. It feeds mostly on plants. Gestation lasts 8–8½ months and the number of offspring varies from one to three. The Spectacled Bear is the only member of the special subfamily Tremarctinae. All other species of bear, including the following one, belong to the subfamily Ursinae.

Polar Bear (201) inhabits the arctic regions. It has never lived in the Antarctic. An excellent swimmer, it is dependent on the presence of various species of pinnipeds and on open water. In addition to fish and pinnipeds, its staple food, the Polar Bear feeds on sea birds, and sometimes hunts even reindeer, musk deer, arctic hares, etc. In summer, it will not despise various fruits and leaves of tundra plants, and occasionally eats seaweeds. It does not usually hibernate, but pregnant females always build deep dens in the snow, giving birth to cubs and falling into torpidity. Gestation lasts about 9 months; usually two young are born. Polar Bears are among the largest bears in the world: their weight reaches 700 kg and they may measure up to 2.25 metres.

Thalarctos maritimus
Ursidae

African Civet (202) is a typical member of the civet family, a phylogenetically ancient group, indicating to a certain extent the appearance of original carnivores in the early Tertiary. Civets are mostly small animals. Some of them have retractile claws. Their characteristic feature is the presence of scent

Viverra civetta
Viverridae

211

212

glands in the anal region. The secretion of scent glands of some species — including the African Civet — is used in medicine (the so-called vivereum), and in the perfume industry. The African Civet occurs from Senegal and Somalia to southern Africa. It measures 110—120 cm including the tail, which accounts for one-third of the length. The body weight is 6—8 kg. Civets eat chiefly small vertebrates, occasionally plants. After 12—15 weeks of gestation, the female deposits two or three young.

Common Genet (203) has semi-retractile claws. The adult reaches a total length of 1 metre; half of this is the tail. The weight is 1—2 kg. It is resident in south-western Europe, a large part of Africa and in the Near East. Its diet consists mostly of insects or other invertebrates, and of small vertebrates. Gestation is 10—12 weeks; usually two or three young are born.
Genetta genetta
Viverridae

Palm Civet (204) grows up to 120—130 cm in length, slightly less than a half of that being the tail. Its usual weight is 2.5—3 kg, but it may reach as much as 4.5 kg. Colouration is greyish to brown, with dark spots. Its area of distribution covers India, Sri Lanka, southern China, all of south-eastern Asia, the Sunda Islands, Borneo (Kalimantan) and the Philippines. The diet is mixed but consists primarily of flesh. The Palm Civet occurs often near human dwellings, hunting mice and rats, sometimes even poultry. Females usually have two litters a year, averaging two to four young.
Paradoxurus hermaphroditus
Viverridae

Binturong (205) is the largest civet. It is native to Nepal, Sikkim, Bhutan, Assam, Burma and all of south-eastern Asia. Its body length averages 70—95 cm; the length of the tail is 60—90 cm. The weight reaches 10—15 kg. The Binturong's hair is long and dark-coloured, with silvery tips. It is mostly nocturnal
Arctictis binturong
Viverridae

213

and prefers living in trees. It is a remarkable climber and its prehensile tail even enables the animal to hang from branches. The teeth are rather weak, since the Binturong eats mainly fruits and tender parts of plants, and only occasionally insects, small vertebrates and carrion. After 90–92 days of gestation, one or two young are born.

214

215

African Mongoose (206) is brown-grey or olive-grey, and reaches a total length of 1 metre, the tail being a little less than a half of that. It is resident in the Iberian Peninsula and the larger part of Africa, except deserts and semi-deserts. In Ancient Egypt, it used to be worshipped as the so-called 'pharaoh's rat', and embalmed bodies of mongooses were deposited in tombs and temples. The African Mongoose feeds chiefly on birds and eggs, but also on snakes, even on venomous species. Gestation lasts about 12 weeks; the litter averages two or three young.

Herpestes ichneumon
Viverridae

Common Mongoose or **Newara** (207) is the famous Rikki-Tikki-Tavi whose fight with a cobra was quite correctly described by Rudyard Kipling in his *Jungle Book*. Like the preceding species, the Common Mongoose also kills snakes, in

Herpestes edwardsi
Viverridae

216

217

addition to small mammals and birds. Neither species, however, is immune against the venom; they win the fight thanks to their agility and speed. The Common Mongoose occurs from western Assam over India, Pakistan, Afghanistan, Iran, Iraq, to the eastern and southern coast of the Arabian Peninsula. It is slightly smaller than the African Mongoose. Gestation lasts 8–9 weeks; the litter averages two to four young. Because of its ability to hunt venomous snakes, the Newara was successfully introduced to the Malayan Peninsula. It was also brought as a rat-killer to many of the islands of the Antilles where, unfortunately, it took to killing poultry, and also began spreading rabies. This example demonstrates that introduction of new animal species cannot be carried out without thorough research and consideration of all the circumstances.

218

219

Banded Mongoose (208) has an attractive brown-grey coat. It measures 60 – 75 cm, of which the tail accounts for 25 – 30 cm. It frequents stony, bushy regions southwards from the Sahara to the Orange River. It lives in groups of ten to 20 individuals, which never stay long in one place. Gestation takes 8 – 10 weeks; three or four, exceptionally even six young are born.

Mungos mungo
Viverridae

220

221

Suricata (209) is widespread in southern Africa, mainly southwards from the Orange River.
Suricata suricata
Viverridae

With the 18–25 cm long tail, it measures 50–60 cm. Its fur is soft, with relatively long hair. Suricatas live usually in colonies, and feed on insects and small vertebrates. They produce two to four young.

Fanaloka (210) lives in the virgin forests and bush of Madagascar. Its colouration is greyish to
Fossa fossa
Viverridae

grey-brown, with a pattern of dark spots which are less distinct on the front part of the body. The uderside is light-coloured. The Fanaloka measures 60–65 cm including the tail, which is some 20–25 cm long. It hunts mostly small verterbrates but will take also various fruits. Thanks to its scientific name, it is sometimes mistaken for the somewhat larger Fosa (*Cryptoprocta ferox*), which also inhabits Madagascar. A number of features of the skull distinguish the Fosa from other civets, and therefore it is sometimes classified in a separate family, Cryptoproctidae.

Aardwolf (211) is the only member of the highly specialized family of aardwolves. It resembles
Proteles cristatus
Protelidae

hyaenas in the shape of its body, but its skull is more elongated, the face has a different structure, and the whole skeleton is more slender. The longer front limbs have five digits, hind feet have four; hyaenas have five digits on both front and hind limbs. Difference is also discernible in the

Satiated young fennec settles down to sleep in the mother's fur: only the front part of the head shows

222

structure of the teeth: aardwolves have small teeth, which are widely spaced, the molars being completely rudimentary. This is related to the nature of their diet, based on insects (mostly termites), exceptionally supplemented by birds' eggs. The Aardwolf is coloured yellowish grey, with dark, crosswise stripes. It has a conspicuous mane of long hair, stretching from the nape of the neck over the back and continuing in the bushy tail. This mainly nocturnal carnivore is native to the arid and open bushy or stony and rocky regions of southern and eastern Africa, from the Cape Coast to Sudan.

Spotted Hyaena (212)
Crocuta crocuta
Hyaenidae

is the largest of the three species of the family. Like aardwolves, hyaenas have front legs somewhat longer than the hind ones, and the front of the body is more massive. All hyaenas prefer a nocturnal life. Hyaenas do not feed only on carrion, but also actively catch live prey (mainly sick and weak animals). They have powerful teeth and can crush even the large, strong bones of large hoofed mammals. Hyaenas emit a range of sounds: the most notorious is their maniacal cry, characteristic of the Spotted Hyaena. The Spotted Hyaena is distributed over the steppe and

223

bushy regions throughout almost all Africa south of the Sahara. The adult reaches 150—190 cm in length, of which the tail constitutes 25 to 40 cm. The height at the shoulder varies between 75 and 90 cm, the weight is 50—82 kg. Spotted hyaenas live in packs of five to 15 animals. Gestation takes about 110 days; one or two young are born.

224

225

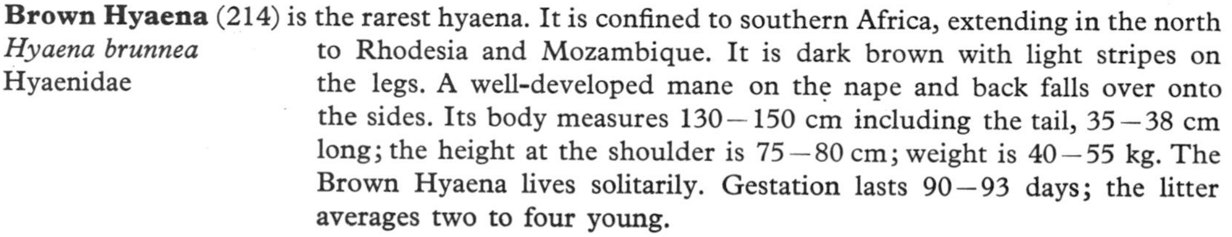

Striped Hyaena (213) is smaller than the Spotted Hyaena. It lives in western, northern and eastern Africa, except primary forests and deserts. It is also found in western and central Asia, in Asia Minor, in Afghanistan, Pakistan and India. Colouration is light or whitish grey with dark or black stripes. Longer hair on the nape and back forms a more or less conspicuous mane. The Striped Hyaena lives singly or in pairs, also in family packs. Gestation is about 90 days; usually two to four young are born.
Hyaena hyaena
Hyaenidae

Brown Hyaena (214) is the rarest hyaena. It is confined to southern Africa, extending in the north to Rhodesia and Mozambique. It is dark brown with light stripes on the legs. A well-developed mane on the nape and back falls over onto the sides. Its body measures 130 – 150 cm including the tail, 35 – 38 cm long; the height at the shoulder is 75 – 80 cm; weight is 40 – 55 kg. The Brown Hyaena lives solitarily. Gestation lasts 90 – 93 days; the litter averages two to four young.
Hyaena brunnea
Hyaenidae

226

227

Dogs (Canidae) are a well-known family of carnivorous mammals. The majority of species in this rather numerous family have a long, slender head, relatively long limbs with, except in one species, five digits on the front feet and four on the hind ones. Their claws are strong, blunt and non-retractile. Ears are large and erect; the tail is well-developed. Smell is their most acute sense and they belong among the mammals with the most superior sense of smell. Most dogs live in family groups, often gathering in small or large packs. Their diet consists chiefly of flesh, and to a certain extent of plant food. This is reflected in their dentition, which has some features of omnivorous mammals, such as comparatively large incisors and broad molars, but is also suited to a predatory way of life in the prominent canine and carnassial teeth.

Dingo (215)
Canis dingo
Canidae

is a wild descendant of domesticated dogs, taken to Australia long before the arrival of white men. Over a long period of time, their appearance settled into one pattern, and a new species of wild dog developed secondarily. The Dingo measures 70—95 cm, tail 30—35 cm; height at the shoulder is 55—70 cm; weight is 15—25 kg. It lives in family groups or small packs, in which it hunts small and bigger vertebrates; larger prey is usually chased to death. Gestation lasts about 63 days, as in the domestic dog; three to six young are dropped. In 1954, a wild dog related to the Dingo and named the 'Forest Dingo' was discovered in New Guinea. It was considered to be a breed different from true dingoes. It is therefore introduced here as an independent species, *Canis hallstromi*, as its taxonomic status is not yet quite clear.

228

Wolf (216), the uncontested ancestor of all breeds of domesticated dog, is the best-known canine
Canis lupus
Canidae

beast of prey, for centuries figuring in many tales of the imagination. The body of adults is 100—140 cm long, the tail measures 35—40 cm; height at the shoulder is 65—85 cm and the weight is 30—75 kg. Males are larger than females. The Wolf was distributed originally throughout Europe, and a large part of Asia and North America. Today, it has been exterminated in vast territories. It lives in families, gathering in the autumn into packs led by the strongest male. Packs hunt large hoofed animals, but they are often satisfied with rodents and birds. Cases of wolves attacking men are extremely rare, and their plausibility is not always reliable. Gestation lasts 63—65 days; the litter averages four to six young.

Common Jackal (217) is much smaller than the Wolf; it weighs 7—14 kg. It is resident in south-
Canis aureus
Canidae

eastern Europe, north and east Africa, western and central Asia, and northern India. Its diet consists of carrion, small vertebrates and fish, and fruits. Gestation lasts 62 days; usually three to eight young are born.

Black-backed Jackal (218) differs from the Common Jackal in its reddish yellow-brown colouring,
Canis mesomelas
Canidae

with a distinct stripe of brown-black and grey-black hair on the back. It inhabits the African steppes and open, bushy areas from Nubia, the Sudan and Somalia to the Cape Province coast. Gestation takes 60 days; the litter numbers five to seven, or even nine young.

Red or **Common Fox** (219) is found almost everywhere in Europe, northern and central Asia,
Vulpes vulpes
Canidae

and North America. It settles in various types of environment; even in the outskirts of towns. The body size and colour vary, but most forms are of typical reddish brown or pale reddish shades. The Common Fox feeds chiefly on small mammals (mainly rodents), on birds, invertebrates,

229

and berries. Gestation lasts about 52 days; usually three to six, but sometimes as many as nine young are born.

Arctic Fox (220) lives in circumpolar regions, and in tundras of northern Asia and North America.
Alopex lagopus
Canidae

It is somewhat smaller than the Common Fox. The Arctic Fox exists

230

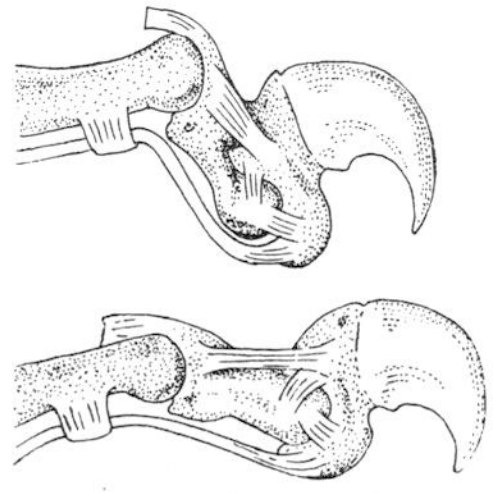

Diagram of retractile claws of feline carnivores

231

in two colour phases: one is snow-white in winter and pale tawny-grey in summer; the second bluish grey or grey-brown in both summer and winter. The second colour phase represents the well-known but relatively rare 'Blue Fox'. The Arctic Fox feeds on small northern rodents, fish thrown up on the shore, insects, remains of the prey of Polar Bears, etc. The gestation period lasts 50—52 days; the typical number of offspring is four to five, but sometimes as many as ten are born.

Korsak Fox (221) is about the same size as the Arctic Fox. It is an inhabitant of the steppe and semi-desert regions from the lowlands east of the Volga to Mongolia and Manchuria. Its diet consists of small rodents, insects and other invertebrates. Gestation lasts 49—51 days; litters average two to four, exceptionally even eleven young.

Alopex corsac
Canidae

232

233

Fennec (222) inhabits the dry steppes and semi-deserts of northern Africa, the Sinai Peninsula, and Arabia. The Fennec is 53—70 cm long, including the tail of 18 to 28 cm. The ears reach 10—12 cm in length. The body weighs roughly 1 kg, rarely 1.5 kg. The Fennec lives on small rodents, birds and their eggs, but mostly on insects. Gestation lasts 49—50 days; two to four, exceptionally five young are produced.

Fennecus zerda
Canidae

Raccoon Dog (223) remotely resembles true raccoons. It is indigenous to the Far East: the Amur region, Manchuria, Korea, China, and Japan. Because of its valuable fur, it was introduced into the European part of the former USSR, and it has been spreading from there in recent years to central and western Europe. Its length varies between 65 and 80 cm, including the tail, which is 15–23 cm long; height at the shoulder is 23–25 cm and the animal weighs 5—9 kg. The Raccoon Dog lives on small vertebrates, birds' eggs and insects. It is the only canine beast of prey which hibernates. Gestation lasts 59—71 days; a typical litter numbers six or seven young, but cases of 16 cubs in a litter have been reported.

Nyctereutes procyonoides
Canidae

Maned Wolf (224) is distinguished by its extremely long legs. The body reaches 120 cm in length and the tail about 40 cm, the height at the shoulder is up to 85 cm. Its weight may be as much as 20—25 kg, exceptionally even more. The Maned Wolf is resident in open or lightly wooded savanna of South America, from central Brazil over eastern Bolivia and Paraguay to northern Argentina. It feeds on small rodents, birds and their eggs, also on fruits of some plants; only exceptionally it may attack sheep. Gestation lasts 62—65 days; the female usually gives birth to two or three, exceptionally even five young.

Chrysocyon brachyurus
Canidae

234

Bush Dog (225) frequents mainly wooded and shrub-covered regions, less often open savanna.
Speothos venaticus
Canidae

It is found from Panama to Brazil, northern Bolivia and Paraguay. Its appearance is reminiscent of mustelids (weasels). It measures 70—90 cm with the tail, which is only 12—15 cm long. It weighs 5—7 kg. Bush Dogs live in family packs, hunting mainly rodents, birds, etc. Packs follow their prey even to water; of all the dogs, they are the best swimmers. Four to six young are born, after a gestation of 2 months.

African Big-eared Fox (226) is a fawn-coloured inhabitant of arid steppes in eastern and southern
Otocyon megalotis
Canidae

Africa. The animal weighs 3—4 kg and reaches 90 cm in length, of which the tail measures some 35 cm; the height at the shoulder is 40 cm at most. The remarkably large, pointed, black-tipped ears may be 13 cm long. This carnivore is endowed with the greatest number of teeth of all beasts of prey: 48 (each half of each jaw contains three incisors, one canine tooth, four premolars, four molars); the teeth are, however, rather small and weak. Its diet is composed chiefly of insects, (predominantly termites), small rodents, birds, and parts of plants. Gestation lasts 60—70 days; two to five young are born.

Asiatic Red Wolf or **Dhole** (227) inhabits mostly forests or forest-steppes from southern Siberia,
Cuon alpinus
Canidae

the Amur region, Korea, central, eastern and southern China to western and eastern India, Sumatra and Java. In various parts of its territory, it is called the Red Wolf, Dhole, Colsun, Adjag, or Serigala. The body measures 75—100 cm, the tail 30—48 cm; the weight is 14—23 kg. The Red Wolf lives in family groups or packs of up to 30 individuals. Packs daringly attack even larger prey such as bears, buffaloes, gaurs and bantengs (two kinds of Indian wild oxen), and a pack is known to have killed a Leopard, and even a Tiger. They never attack Man. Solitary or paired Red Wolves eat just small rodents, birds, etc. Gestation takes 48—50 days; the litter averages two to six.

African Wild Dog (228) ranges southwards from the Sahara to the Cape of Good Hope. It can be encountered in the savanna, as well as in eastern African mountains, to the upper tree line. It is of the same size or bigger than the Asiatic Red Wolf. It is the only canine beast of prey with four toes on both front and hind limbs. Colouration consists of irregular, white, black, and ochre-brown spots. The African Wild Dog lives in packs of four to 60; packs hunt hoofed mammals, chiefly antelopes. Gestation lasts 63 to 72 days; the litter numbers five to eight young.

Lycaon pictus
Canidae

Cats (Felidae) are the carnivores best-adapted for hunting live prey. They are the most perfect carnivorous mammals, differing from the others in a more rounded shape of the head, shorter, but stronger jaws (in this respect, only hyaenas surpass them), and in fewer teeth (most have only 30 teeth), which are, however, powerful, with perfectly and strongly developed canine and carnassial teeth, serving as the main tools for grasping, killing and tearing up the prey. The body is sinewy but slender and flexible; the limbs are moderately long, as is (usually) the tail. Save two exceptions, all cats have completely retractile claws, which, when the animal is tranquil, are hidden in a ligamentous sheath and fur and are not visible. The razor-sharp, sickle-shaped claws are excellent instruments in helping the animal to grasp and hold its prey. Female cats are regularly somewhat smaller than males.

235

236

Flat-headed Cat (229) illustrates to a certain extent the appearance of original cat forms. Its short limbs and the elongated, pointed and flat head remind one of the civets. Its claws are not entirely retractile: when retracted they do not touch the ground, but the tips stay out. The Flat-headed Cat lives in the Malayan Peninsula, Sumatra and Borneo (Kalimantan). Its body measures 54—62 cm, the tail is 15—21 cm long. It is a rare species, and almost nothing is known about its habits. It occurs mostly in primary forests and waterside bushes, feeding on small reptiles, amphibians and fish. Occasionally it catches birds, and it also eats fruits of certain plants.

Ictailurus planiceps
Felidae

Leopard Cat (230) is of the size of a domestic cat, but its limbs are slightly longer. Basic colouration is ochreous, sometimes with a reddish overtone, with a pattern of dark to black spots. It is widespread from Pakistan to Cashmir, and to the Philippines and the Greater Sundas. Some authors range this species together with the Amur Forest Cat (*Prionailurus euptilurus*), which is somewhat larger, and thus include Manchuria and Eastern Siberia in the area of distribution of the Leopard Cat. Gestation lasts 65—70 days; usually two to four kittens are born.

Prionailurus bengalensis
Felidae

Ocelot Cat (231), sometimes called Little-spotted Cat or Oncilla, is a relatively little-known species, native to South American primary forests and bush. It is about the same size as the Leopard Cat, but the tail is somewhat longer. Its body weight is 1.5—3 kg. The diet consists of small vertebrates. Females undergo 74—76 days of gestation, and bear one or two young.

Leopardus tigrinus
Felidae

Tree Ocelot or **Margay** (232) lives from Mexico and Panama to Uruguay and northern Argentina.
Leopardus wiedi
Felidae

It is almost exclusively a primary forest species, climbing trees with extreme agility. Its body measures 50–75 cm, the tail 32–52 cm. It feeds mainly on small birds and mammals. Gestation takes approximately 70 days; the litter size varies between two and three young.

Pampas Cat (233) is the most common South American cat, distributed both in open savanna
Lynchailurus colocola
Felidae

and in bushy and forest areas throughout South America. Its colouration ranges from whitish-fawn tones to tawny-grey, always with an outstanding pattern of dark spots and stripes. The long fur often forms a mane on the back. The body measures 56–70 cm, the tail is 25–30 cm long. The Pampas Cat feeds on small mammals and birds. The probable duration of gestation is about 67 days; one to three kittens are born.

Jaguarondi or **Eyra** (234) has an elongated body, up to 75 cm long, on relatively short legs. The
Herpailurus yagouaroundi
Felidae

tail reaches 60 cm; the weight is 9 kg at the most. The Eyra occurs from Arizona and Texas to southern Brazil and northern Argentina. It prefers thickets and dense primary forests, where it catches mainly birds and small mammals. It is a good climber. Gestation lasts about 70 days; two to four kittens are born.

European Wild Cat (235, 236) is the ancestor of the domestic cat, but it is stouter. Old males
Felis silvestris
Felidae

weigh up to 8 kg; reports on wild cats exceeding 15 kg in weight are exaggerated. The European Wild Cat ranges almost all over Africa and Europe, and a large part of Asia (in many places, particularly in Europe, has it been exterminated). Within this vast range of distribution, numerous geographical forms are recognized: the best known one, living in European forests, is the European Wild Cat (*Felis silvestris silvestris*) (235).

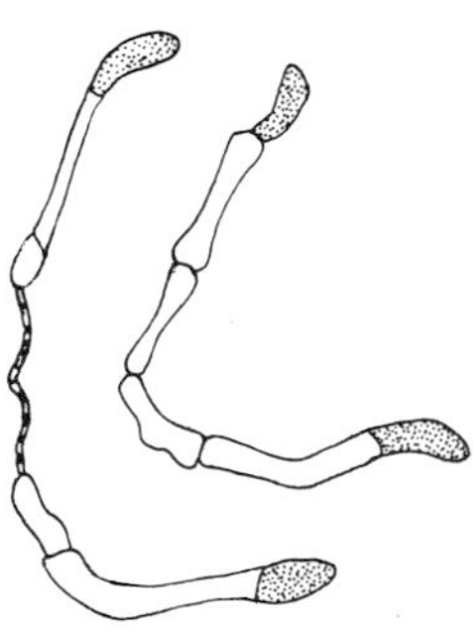

Differences in the arrangement of the hyoid bones in small cats (right) and big cats, in which bones are connected by a flexible ligament (left). This difference is one of many features distinguishing the two groups

237

238

Steppe forms include the Indian Desert Cat (*Felis silvestris ornata*) (236). Pattern and colouration of the individual forms varies but tawny, tawny-grey or grey tones always prevail. Wild cats feed chiefly on small vertebrates. Gestation lasts on the average 66 days; the litter consists most often of two to four kittens.

Jungle Cat (237) is a grey-yellow coloured inhabitant of north-eastern Africa (Nile River basin), western and central Asia, Asia Minor, India, and Sri Lanka. It lives mostly in reeds and bushes on banks of rivers and lakes. It measures 82—105 cm, including the 23—29 cm tail: it weighs 5—11 kg. Its diet chiefly includes smaller mammals and birds. Following 65—67 days of gestation, the female bears three to five young.

Felis chaus
Felidae

African Desert Cat (238) lives in the most arid, semi-desert or desert regions of north Africa, Arabia and central Asia. Its body measures 45—57 cm, the tail is 28 to 35 cm long. It feeds mostly on small desert vertebrates and insects. Gestation lasts approximately 60 days; two to four young are born.

Felis margarita
Felidae

Pallas' Cat (239) frequents steppe, bushy and rocky mountainous locations in Asia, where it occurs discontinuously from the Caspian Sea to the Trans-Baïkal region, and south to Iran, Afghanistan and Tibet. The very long, yellowish grey fur is adorned with a distinct pattern of dark stripes and spots. The body measures 50—65 cm, the tail 21—32 cm; the animal weighs 2.5—4 kg. Pallas' Cat feeds on various smaller vertebrates. Gestation lasts 60—65 days; two to four kittens are born.

Felis manul
Felidae

239

Serval (240) is a long-legged cat, widespread over almost all of Africa except desert regions. It prefers dense vegetation: tall steppe grass, bush or forest borders. Its body is 65–90 cm long, and the height at the shoulder is 40–65 cm. The Serval preys chiefly on rodents and medium-sized birds, only rarely on small antelopes. Gestation takes 68–74 days; litters consist of 2–4 young.

Laptailurus serval
Felidae

240

241

242

243

Northern Lynx (241) has been exterminated in many places, although it was originally distributed
Lynx lynx
Felidae

over almost all of Europe, large parts of Asia, and the northern part of North America. Its characteristics include tufts of long hair on the ears, conspicuous whiskers, a very short tail, and relatively long, stout legs. The height at the shoulder is 60–75 cm, body length 80–125 cm, the tail is 11–23 cm long; the weight is 15–38 kg. The Lynx inhabits dense forests and impenetrable mountain valleys and gorges. Its diet includes small and medium-sized vertebrates, from amphibians, mice and voles to hares and small hoofed mammals. It rarely kills bigger animals. The Lynx is an excellent tree-climber, and it can leap as far as 8 metres. Gestation takes 65–70 days; two or three young are born.

Bobcat or **Bay Lynx** (242) lives in North America, from Mexico to the southernmost part of
Lynx rufus
Felidae

Canada. It is greyish beige, with a marked pattern of small black or dark brown spots. Old specimens, 90 cm long and 58 cm high at the shoulder, do not weigh more than 15–16 kg. The tail may be up to 19 cm long. The Bobcat inhabits not only forests, but also semi-desert areas. Its food consists of smaller mammals and birds. Gestation lasts 60–70 days; two to four young are born.

Caracal or **Desert Lynx** (243) is not a lynx in the true sense of the word. It has a longer tail, no
Caracal caracal
Felidae

whiskers, and its colouring is a uniform reddish ochre or fawn. Its body measures 65–80 cm, the tail 22–32 cm, and its shoulder height is 42–48 cm. The Caracal lives in steppes, bushy or semi-desert areas of all Africa, central Asia, Asia Minor, Iraq, Afghanistan, and India. It preys on small mammals, birds up to the size of a pheasant, and often on small reptiles. After some 70 days of gestation, the female drops two or three young.

244

African Tiger Cat or **African Golden Cat** (244) is found in the forests of central Africa. It stands about 50 cm at the shoulder. Its body measures 65—85 cm and the tail is 27—40 cm long. Large specimens weigh about 15 kg. Colouration varies between chestnut, grey-brown and silvery grey shades. Some specimens are completely covered by small brown or dark-grey spots, while others have spots only on the underparts and insides of the limbs. The diet consists of birds and small and medium-sized mammals. The period of gestation probably lasts some 70 days. The litter numbers one to three young.

Profelis aurata
Felidae

245

Indian Golden Cat (245) is a somewhat larger Asiatic relative of the previous species, also confined to life in primary forests. The colour of its coat varies from golden and reddish brown to pale grey-brown; the cheeks are marked by a narrow white stripe, lined with black. Entirely black (melanic) phases also occur. Its diet is formed of small and medium-sized mammals, although the Asiatic Golden Cat is said to be capable even of killing calves of domesticated buffaloes. In duration of gestation and number of offspring, it probably does not differ from the African Tiger Cat.

Profelis temmincki
Felidae

Marbled Cat (246) is one of the most beautiful and rarest species of the cat family. It lives in Nepal, Sikkim, Burma, Indo-China, Malaysia, Borneo, and Sumatra. It measures 48–63 cm, the tail is 35–45 cm long; its weight is about 4–5 kg. The Marbled Cat is a typical inhabitant of dense forests, and it is one of the best climbers among the cats. It hunts mainly birds and small mammals, but also smaller reptiles. The probable duration of gestation is about 65 days; two or three young make up the litter.

Pardofelis marmorata
Felidae

Puma or **Mountain Lion** (247) is a large cat inhabiting virgin forests, bushy grasslands, rock deserts, and river valleys. It is widespread in the western part of North America, and over almost all of South America. Adult males reach 100 to 150 cm in length, the tail measures 65–80 cm; the weight is 45–100 kg. The Puma preys chiefly on medium-sized and large mammals, less frequently on birds. It is noted for its agility when moving on tree branches. Gestation lasts 90–96 days; the number of offspring varies between one and five.

Puma concolor
Felidae

246

247

Clouded Leopard (248) is an extremely nimble and boldly coloured animal. The smooth, glossy, brown to yellow-buff fur is patterned with large, irregular spots on the flanks; each spot shades to the body colour on one side, while on the other side it is darker, and black-edged. The body measures 100—105 cm, the tail up to 85 cm. As a result of the slenderness and length of the body, the weight is only 20—30 kg. The Clouded Leopard is a typical inhabitant of the forests of southern and south-eastern Asia. Since it is an expert climber, it is usually considered an arboreal species feeding predominantly on birds. New observations, however, have shown that this cat seeks its food mostly on the ground, hunting medium-sized hoofed mammals, mainly wild pigs. Females undergo 86—92 days of gestation, and drop one to five young.

Neofelis nebulosa
Felidae

Leopard (167, 249, 250) is a typical example of the so-called 'big cats' (subfamily Pantherinae), differing in certain anatomical, morphological and biological features from all species of the cat family mentioned so far, which form a group of the 'small cats' (subfamily Felinae). The Leopard used to be widely distributed throughout Africa (except the completely desert areas), and over vast areas of Asia including the Sinai Peninsula, Asia Minor, the

Panthera pardus
Felidae

248

Caucasus, central Asia, India, Indo-China, the Malayan Peninsula, a considerable part of China, Sri Lanka, Java and the Kangean Archipelago. Why no traces of leopards were ever found in Sumatra remains a mystery, as this island is a 'bridge' between the Malayan Peninsula and Java. Today the Leopard has been exterminated in many places. The body length of an adult male reaches 100—145 cm, the tail is 70—96 cm long, its height is 60—68 cm; the weight is 40—68 kg. Melanic specimens (250), the so-called 'black panthers', often occur; they still are often erroneously regarded as an individual species. Leopards living far to the north are conspicuous by their tawny-coloured, soft and very long fur. The Leopard feeds on medium-sized and large mammals, less frequently on birds. Like other large cats, it sometimes attacks domesticated animals, and some specimens may even specialize in hunting them. Man-eating leopards occur rarely: the most ill-famed of them was the so-called Rudraprayag Leopard, noted for killing 125 people, according to official records. Leopards are excellent tree-climbers; they often hide their prey in trees (167). Gestation lasts 90—92 days; two or three young are born.

249

Jaguar or **American Tiger** (251) is the largest predatory cat of the New World. It occurs from
Panthera onca
Felidae

the southern part of the United States over Central and South America, to ca. 40°S. Like the leopards, jaguars have been exterminated in many places. In colouration, the Jaguar resembles the Leopard, but the dark spots and rosettes are bigger, often enclosing one or more smaller dots. Melanic specimens also occur in jaguars. The Jaguar is larger and more robust than the Leopard (weight up to 100 or exceptionally even 150 kg). It prefers forests and thickets close to water, but can be found even in open savannas. The Jaguar is a good climber. It preys mainly on larger mammals, and sometimes feeds on large birds and fish. Gestation lasts 93–105 days; litter size is one to four young.

Lion (252, 253, 254), until recently, inhabited almost all of Africa, except the heart of the African
Panthera leo
Felidae

forest; it also lived in south-western Asia, from Asia Minor to western India. Today, it is found in Africa only in wildlife reserves, and on the Asian continent in a single location – on the Kathiawar Peninsula in the Indian state of Gudjarat. Beside the Tiger, the Lion is the largest cat (total length of males is 280–300 cm, weight 180–220 kg). It lives in prides numbering 15 or more individuals. It lives in both open and wooded savannas, where it hunts mostly large ungulates. The Lion can sometimes be seen in trees, but its movement there is clumsy, and it leaves the ground only to rest. Moreover, it has to choose a tree with branches rather near the ground. Some lions specialize in killing, cattle causing damage to the herds of local farmers. A man-eater may occur, but these cases belong mostly to the past. Gestation lasts 105–108 days

In the mating season, males of many feline beasts of prey fight for females

250

251

252

253

The Cape Lion, *Panthera leo melanochaita,* **was the largest lion subspecies, conspicuous for its massive black mane spreading to the abdomen. It lived in central Cape Province and in the Orange State, but was exterminated in the 1850–60s.**

on average and the litter numbers three or four cubs. Melanism in lions is not known, but white lions sometimes occur as an extreme exception; even though the coat is snow-white, it is not a case of true or total albinism, as the colour of eyes is not red, but usually yellowish, the same as in other lions.

Tiger (255, 256) is restricted to the Asian continent, where it used to range from the Caucasus
Panthera tigris
Felidae

and Asia Minor, across central Asia to the Trans-Baïkal region and the Amur-Ussuri area. It also lived in northern Iran and Afghanistan, in India and south-eastern Asia, in China, on the Malayan Peninsula, and on the islands of Sumatra, Java and Bali. It never occurred in Sri Lanka and other Asian islands. Only in the north, it sometimes swam or crossed over ice to the Sakhalin island, just north of Japan. The Tiger has been exterminated in most of its former territories, and it is true to say that the total number of Tigers living in the wild today does not exceed 3,000 to 3,500 specimens.

254

Like other mammals with a large area of distribution, the Tiger has several geographical races or subspecies, differing in size, colouration, arrangement of stripes, length of hair, and some skull characteristics. Large subspecies such as the Indian Tiger (*Panthera tigris tigris*) (256), and especially the Ussurian Tiger (*Panthera tigris altaica*) (255), are bigger than lions and, as the largest living cats, they rank among the largest carnivores of all. Old male Ussurian tigers, conspicuous by their long winter hair and incredible size, attain a total length of 310 – 325 cm, including the tail, which is 100 – 105 cm; they stand 105 cm at the shoulder, and may weigh more than 250 kg. Verified maximum dimensions of a male Ussurian Tiger are: 335.5 cm total length and 110 cm shoulder height. The greatest recorded weight is 306.5 kg. The Tiger is mostly solitary, preferring thick forests, jungles and inaccessible rocky sites. It can even be found at altitudes above 2,500 metres. The Tiger feeds chiefly on wild boars, various species of deer, antelopes, buffaloes, and, in eastern Siberia and Manchuria, even on bears. Tigers sometimes specialize in cattle-killing, and they have become ill-famed for man-eating, although this has always been exaggerated. There were few man-eaters such as the notorious 'Champawat Man-Eater', which had killed 430 people and was shot by the famous naturalist and expert on Tigers Jim Corbett. Gestation lasts 104 – 107 days on average, and females give birth to two or three young. Seven cubs were born in one particularly unusual case. Some reports and observations mention the rare occurrence

255

256

of black Tigers. There is, however, no verified specimen to bear out these accounts. On the other hand, the so-called 'white Tigers of Rewa' are well-known. These Tigers are white-coloured, but they have distinct dark stripes and blue eyes; the eyes of true albinos would be red.

257

Snow Leopard or **Ounce** (257) inhabits the impenetrable valleys and mountain slopes of Central Asia. Its body attains 100–120 cm in length, the tail is 80–90 cm long, shoulder height is up to 60 cm, and the animal weighs 40–60 kg. The Snow Leopard feeds on mountain sheep and ibexes, marmots and birds. It climbs trees with agility, and can jump well: a 15-metre-long jump has been recorded. The female is pregnant for 98–103 days; two or three young are born.

Uncia uncia
Felidae

Cheetah (258) is the only member of the last subfamily of cats (Acinonychinae). It differs from all other cats in the shape of its body, resembling rather that of a greyhound than that of a cat. It has non-retractile, blunt claws, and many other morphological and anatomical peculiarities. It lives in open grassy plains; it used to be distributed from India and central Asia to Africa, where it resided in suitable habitats from the Atlas Mountains to the Cape of Good Hope. Today, it has been almost exterminated in Asia, and is becoming quite rare in Africa. The Cheetah hunts its prey – medium-sized antelopes – on the run. Although it has been proved to be the fastest mammal (reaching a speed of 90 km per hour), it is not a long-distance runner, and if it fails to catch its prey, it has to give up the chase after several hundred metres. In the past, Indians used Cheetahs in ceremonial hunts. Body length of a male Cheetah is 110–130 cm, the tail has 70–85 cm; the shoulder height is 75–82 cm, and the weight is 40–65 kg. Gestation takes 90–95 days and three to six young are born.

Acinonyx jubatus
Felidae

258

259

Chapter 9 CARNIVOROUS MAMMALS OF THE SEA

In the early Tertiary Era, an evolutionary line broke away from the carnivores; its members specialized in aquatic life. Today they are classified in the order Pinnipedia. Their common ancestry with the carnivores has been demonstrated, and their relation to them is evident in some anatomical characteristics. On the other hand, they differ in numerous adaptations to life and locomotion in water: their limbs (flippers) are fin-shaped, the femur and shoulder bone are shortened and are enclosed within the body, so that the movable, projecting part of the limb is supported by the other bones of the limb, starting at the knee and elbow joints. The limbs have five webbed toes. Further adaptations to an aquatic existence are the spindle-shaped, tail-less body, short thick coat, small head with small or no external ears, nostrils and ear orifices which can be closed when necessary by contraction of a muscle, and a thick layer of subcutaneous fat which protects the animal against cold and water pressure. Some pinnipeds dive to a depth of 300 metres and are able to stay underwater for up to 30 minutes. In comparison with other mammals, they have a relatively greater volume of blood, capable of carrying a considerable quantity of oxygen. Before diving, pinnipeds exhale; the body thus becomes heavier underwater, and the empty lungs are under less pressure. Deoxygenated blood concentrates in the inferior vena cava, and oxygenated blood goes mainly to the brain, whilst the supply of oxygen to the other organs is limited. Pinnipeds live offshore from continents and sea islands and in several cases even in lakes. Their food consists of fish, crustaceans and cephalopods which they do not bite into pieces but swallow whole. Their teeth have sharp-pointed crowns, adapted for grasping the prey firmly. Many pinnipeds live at least for a part of the year in groups. In the mating season, each male selects a number of females. Gestation lasts up to a year, and there is often a latent pregnancy. A single cub or two are born on land or ice; they are fairly independent.

260

The skeleton of the hind limb of a pinniped clearly shows adaptation to aquatic life

261

Australian Sea Lion (259) is a representative of the family of sea lions (Otariidae), most closely allied to the original evolutionary types of the group. Sea lions differ from other pinnipeds in one distinct feature: they have external ears. When moving on land, they can shift the hind limbs under the body and walk, even though clumsily. In the mating season, males set up harems numbering up to 40 cows. The Australian Sea Lion is a small species, inhabiting southern coasts of Australia, its islands and Tasmanian shores. When dry, the hair is coloured grey-brown, wet hair appears to be almost black. Body length in old males is 180–200 cm, weight is 200–220 kg. Females measure 150 cm at maximum; their weight usually does not exceed 100 kg. The diet consists of fish, cephalopods and crustaceans. Gestation lasts approximately 330 days.

Arctocephalus doriferus
Otariidae

Californian Sea Lion (260) lives off the Pacific coast of North America, from British Columbia to Nayarit in Mexico, on the islands Tres Marias, in the Galapagos Islands, near both eastern and western Japanese shores, and in some places off the east coast of Korea. Dry fur is yellow-brown to sepia-coloured; when wet, it seems almost black. Body length in adult males reaches 230 cm, in females 170 cm. Weight in males is 250—280 kg, in females about 90—100 kg. The diet is composed chiefly of fish and various cephalopods. Gestation lasts 340—365 days; a single cub is born.

Zalophus californianus
Otariidae

Steller's Sea Lion (261) is the largest sea lion. Old males are 300—350 cm long and weigh 700—1,100 kg; females measure 200—250 cm and weigh 150—300 kg. Dry fur is yellow-brown, wet fur looks almost black. Steller's Sea Lion inhabits the Pacific coast of north-eastern Asia and North America, from North-Korean shores and northern Japan over the Kuril Islands, the Kamchatka and Chukchi Peninsulas, the Bering Strait, the Aleutian Islands, the Alaskan and Canadian coast to California. Its diet comprises various species of fish, cephalopods and, very rarely, small specimens of seals. The well-known rookeries – sites where sea lions gather in assemblies even of several thousand individuals – are formed mainly in the mating season, that is, in late spring and early summer. Gestation lasts 345–365 days; a single cub is born, two are extremely exceptional.

Eumetopias jubata
Otariidae

262

Skull of the Walrus seen from the side

263

Walrus (262, 263) is the only representative of the second pinniped family, the walruses. It has a stout body, naked except for extremely sparse, tough hairs. The upper lip, however, is covered by the characteristic moustache, consisting of some 400 tough, strong hairs. The skin of the Walrus is very thick and wrinkled, sometimes forming conspicuous callosities in front. The hind limbs can be placed under the body as in sea lions, so that the Walrus can 'walk'. The extremely elongated upper canine teeth, protruding far from the mouth, are another remarkable feature: in the males they can reach up to 80 cm in length, in females up to 40 – 50 cm. Walruses are

Odobaenus rosmarus
Odobaenidae

264

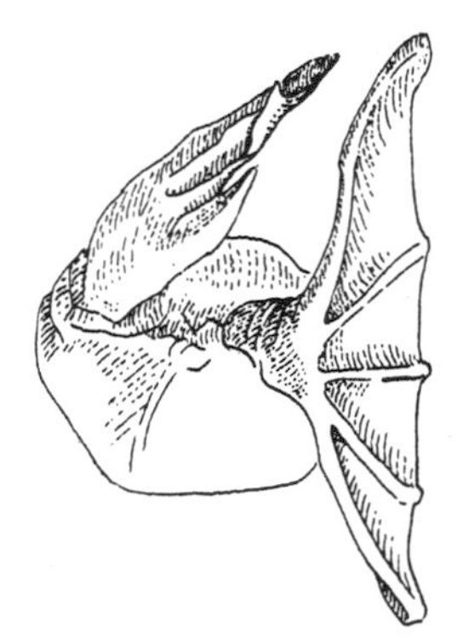

Back view of a swimming walrus, showing the use of limbs modified into flippers for movement and manoeuvring in the aquatic environment

265

bulky animals; they belong to the largest pinnipeds not fearing even polar bears. Old bulls measure up to 350–400 cm and weigh some 1,100–1,500 kg. Females are 330 cm long at the most and they weigh 600–800 kg. The Walrus lives in open waters on ice floes, and off the Arctic Ocean shores, occurring mostly in the northern coast of north-eastern Siberia, north-western Alaska, northern and north-western coast of Greenland, and on the island of Ellesmere. It feeds on gastropods, bivalves, larger crustaceans, occasionally seals, fish, and exceptionally small and medium-sized cetaceans. These last are not killed by the Walrus; it only devours those cast ashore or on ice floes. A walrus family usually consists of the old bull, two to four cows, and several juveniles. Gestation, not latent in this case, lasts 355–360 days, and the female gives birth to a single young.

Harbour Seal (264, 265)
Phoca vitulina
Phocidae

When water freezes over in winter, seals preserve holes in the ice for breathing, even if the ice becomes thick

is a typical representative of the last pinniped family. Of all pinnipeds, seals are the best adapted for life in water. This is seen in the shape of the body and of the flippers, and other morphological features endowing the body with supreme hydrodynamic properties. The hind limbs constitute the main organ of locomotion in water: they are situated at the very end of the body, and act in the same manner as the caudal fin in cetaceans. When seals move on land or ice, the hind limbs cannot be pushed under the body, and when out of water, seals are forced to move in a caterpillar-like fashion – contracting and stretching their bodies. The body has a coat of short, tough hair, and the upper lip is covered by long, tough hairs. The colour of the coat is greyish-white to dark grey-brown, with patterns of either dark or pale spots and rings. Seals live mainly on fish, cephalopods, and some marine invertebrates. Among

266

267

268

their enemies rank some predatory cetaceans, for example, killer whales and, in the Arctic, polar bears. The majority of species of seal stay in pairs during the breeding season, except the Grey Seal and the South Atlantic Elephant Seal, in which males gather harems of cows. The Harbour Seal is widespread from subtropical to subpolar zones of the eastern and western shores of North America and southern Greenland, the Icelandic waters, the western coast of Europe, and the eastern coast of Asia. Males grow up to 170–190 cm in length, and weigh up to 100 kg. Females are slightly smaller. Gestation lasts about 280 days; one or two young are born.

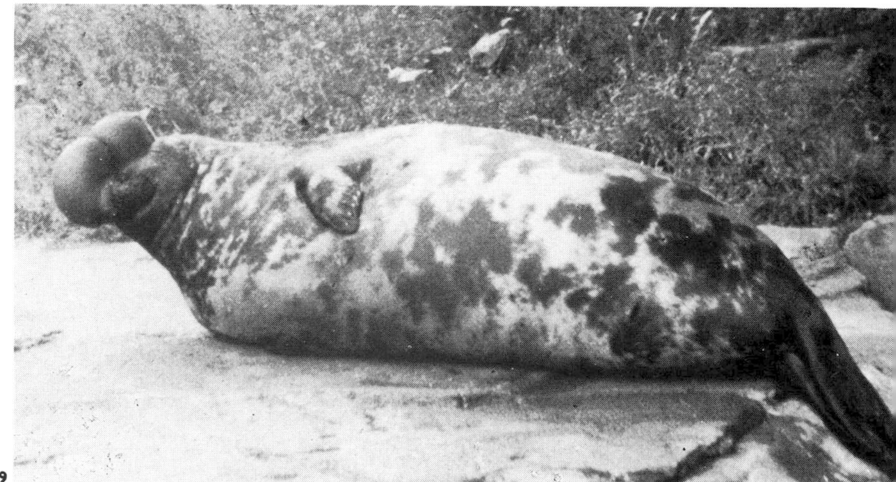

269

Baïkal Ringed Seal (266) lives in the Baïkal, and is the only truly fresh-water pinniped (there is, however, one form of the Harbour Seal which lives in fresh waters of the Lower Seal Lake, situated some 150 km from the Hudson Bay). The Baïkal Ringed Seal is one of the smallest pinnipeds: its body reaches a mere 130—140 cm; its weight is 80—90 kg. Colouration is a uniform brown or brownish silvery grey, slightly paler below. Gestation lasts 325—335 days; usually a single cub is dropped.

Pusa sibirica
Phocidae

Hawaiian Seal (267) is the only truly tropical pinniped species: it lives off the shores of western Hawaiian and Laysan Islands. The upper parts of the body are coloured a uniform brown, the belly is slightly paler, with yellowish spots. Both males and females attain a length of 210 cm at the most, and they weigh up to 150 kg. The probable duration of pregnancy is 300—330 days; a single young is born.

Monachus schauinslandi
Phocidae

Grey Seal (268) inhabits coasts and islands of northern parts of North America, Nova Scotia and to the northern tip of Newfoundland, the European coasts north from Brittany, the Icelandic shore, the coast and peninsula of Scandinavia including the Baltic, and the Kola Peninsula. This species is distinguished by a pronounced difference in size between males and females. Males reach a length of 280 cm and weight of 315 kg, while females measure 220 cm at the most, and weigh only up to 200 kg. Colouration is highly variable, from light grey to almost black, with patterns of irregular, dark or pale spots. Gestation lasts 320—335 days; the female bears one or two young.

Halichoerus grypus
Phocidae

270

271

Hooded Seal (269) is a remarkable resident of the northern Atlantic and Arctic Oceans. Most of the year it travels through the sea and along the coast, staying for a longer period in one place only in the breeding season. Males measure up to 300 cm, females only 230 cm. Maximum weight in males is 410 kg, in females 230 kg. Adult bulls have a conspicuous skin pouch, stretching from the nostrils to the back of the head. When excited, they blow it up with exhaled air, into the form of a cap or hood. Ground colouration of the skin is bluish to dark grey, with paler spots of various shape and size. Gestation lasts approximately 335—340 days; the female bears a single cub.

Cystophora cristata
Phocidae

South Atlantic Elephant Seal (270, 271) is the largest pinniped. Old males may reach up to 650 cm in length and more than 3,500 kg in weight. In females, however, the body length does not exceed 350 cm, and they weigh 900 kg at most. This bluish-grey giant lives off the Antarctic shores, and in the southern Atlantic and Pacific Oceans. Old males have at the front of the head a kind of 'trunk', more than 30 cm long, which they inflate and extend with exhaled air when they are excited. The Elephant Seal's diet consists mainly of fish and cephalopods, as in the majority of pinnipeds. Gestation lasts 345—355 days and a single cub is born; twins are extremely rare.

Mirounga leonina
Phocidae

272

Chapter 10 HORSES AND THEIR ALLIES

If one stops to imagine the Indian One-horned Rhinoceros in full stampede, it is difficult to avoid thinking of it as the nearest thing in the animal world to a living tank. This gigantic animal is classed in the order of perissodactyls (Perissodactyla), including only large mammals, usually perfectly adapted for running. This group reached its evolutionary peak in the Tertiary period, when members of at least 15 perissodactyl families existed. Only three of them have survived till the present time: rhinoceroses, tapirs, and horses. One of the common features of all perissodactyls is the 'mesaxonic' limb, in which, besides a possible reduction of one or more other toes, the long axis passes through the third, middle toe. The third toe is always the strongest, bearing most of the weight and forces on the limb. At their tips, the toes are modified into hoofs, made of the same tissue as claws and nails in other mammals; the hoof covers the end of the last segment of the digit. Perissodactyls walk only on the toes — they are digitigrade — and all of them are herbivorous. They are guided by hearing, smell and sight, their fine sense of hearing and smell, combined with the ability to run quickly, being often their only defence against carnivorous predators. Perissodactyls bear regularly one, very rarely two young. The existence of almost all perissodactyl species is endangered, and some forms (for example the Quagga and the Syrian Wild Ass) have been totally exterminated by Man.

Indian One-horned Rhinoceros (272, 273) is the typical representative of the rhinoceros family.
Rhinoceros unicornis
Rhinocerotidae

It has three-toed feet, and a remarkably thick and deeply folded skin which resembles plates of armour. The body is usually naked, except for several sparse hairs on the ears, and bristles on the tail. Colouration is brownish grey to dark grey. There is a single horn, relatively slender and pointed, generally not exceeding 20—25 cm, although exceptionally it can measure more than 40 cm; the maximum recorded length of a horn

273

274

of an Indian One-horned Rhinoceros is 60.7 cm. Horns are formed from skin, without skeletal support: they adhere to the rough surface of nasal bones. General opinion is that the African White Rhinoceros is the largest species of rhino, but examination of records shows that the Indian species reaches at least the same size. Old males measure 360—400 cm, the tail is 60—75 cm long, and the height at the shoulder is 175—188 cm. The body weight can reach 3,900 kg, although it usually averages 2,000 to 3,000 kg. With the White Rhinoceros, it is the second largest terrestrial mammal after the two species of elephant. Of the five modern species of rhinoceros, three live in Southern Asia, and two in Africa. The Indian One-horned Rhinoceros was originally distributed only in the grassy jungles of aluvial plains in northern India and Nepal, where it occurred

275

276

from the middle Ganges to Assam. Today, it is almost extinct. It has survived in several places: the Kaziranga National Park is the best-known location. The Indian One-horned Rhinoceros feeds exclusively by grazing in grasslands; it is usually solitary, except during the mating season and the time when the young is born. Gestation lasts 465–490 days.

Sumatran Rhinoceros
Didermocerus sumatrensis
Rhinocerotidae

(274) is the smallest species of rhinoceros. Its body length is 240–275 cm, the tail is about 50 cm long, and the shoulder height is 120–140 cm. It weighs up to 1,000 kg. It is the only rhinoceros with significant, red-brown to black hair cover, although this becomes more sparse with age. The skin is greyish to grey-black. The Sumatran Rhinoceros has two horns: the front one is 15–20 cm long, the other measures only 5–8 cm. The maximum recorded length of the front horn is 38.1 cm. The Sumatran Rhinoceros was originally widespread in Assam, Burma, almost all Indo-China, on the Malayan Peninsula, Sumatra, and Borneo. At present, it is nearly extinct; the last few dozens survive in the most impenetrable areas of western Indo-China. It is confined to dense thickets and forests, feeding on leaves and shoots of various bushes and trees. Gestation is estimated, according to unverified data, at 210–240 days, although it will probably prove to be longer – around 440–450 days.

Black Rhinoceros
Diceros bicornis
Rhinocerotidae

(275) is nowadays probably the best known species of rhinoceros. It is indigenous to the African bushy savannas or the dense, thorny bush country from the Cape of Good Hope to Sudan, Somalia, and Lake Chad. Its skin is coloured yellowish grey-brown to dark grey. There are two horns on the head: the front one is longer, measuring usually 40–50 cm, 135 cm at the most. Body length varies between 300 and 350 cm, the tail is 65–80 cm long; the height at the shoulder is 140–165 cm. The Black

277

Rhinoceros weighs 1,500 – 1,800 kg, exceptionally 2,000 kg. The head is rounded in front; the upper lip, unlike the following species, is protracted into a pointed, finger-like outgrowth which helps the animal in breaking and tearing shoots, twigs and leaves. The gestation period lasts 450 – 480 days: a single young is born.

White or **Square-lipped Rhinoceros** (276, 277) ranks among the largest rhinoceroses. Its body measures 340 – 380 cm, the tail is 70 – 90 cm long, the height is 170 to 185 cm. Its weight varies from 2,000 to 3,000 kg, exceptionally 3,600 kg. The colouration of the skin is light grey. Of its two horns, the front one is longer, 50 – 60, often 100 cm; the maximum recorded length is 165.5 cm. The shorter horn measures merely 20 – 30 cm. The White Rhinoceros differs from the previous species, having a wide, blunt mouth with a flat upper lip; a ligamentous hump in the back of the neck, a longer head, and a somewhat different body shape. It feeds on grass, and frequents grass-covered plains and bushy steppes. It is found in southern Africa between the Orange and Zambezi rivers, in northern central Africa, in the Sudan west of the Nile, in Uganda, north-eastern Congo, and in several adjacent regions. It is very rare everywhere. The female gives birth to her single young probably after 470 – 500 days.

Ceratotherium simum
Rhinocerotidae

The second family of perissodactyls are the tapirs. Today, only four species of tapirs exist: three in Central and South America, and one in south-eastern Asia. Tapirs have four toes on the front feet; the external toe is stunted, and does not touch the ground when the animal walks; the hind feet have only three toes. The head has a short proboscis; the tail is short and stumpy. Tapirs can be found mostly in places with an abundant water supply: in virgin forests, savannas, bushy jungles, mountainous forests and bush. They are good runners, swimmers and divers. They live solitarily or in pairs, feeding on aquatic vegetation, leaves, shoots, fruits and twigs. Gestation varies between 390 and 405 days; the litter almost always comprises a single calf. Tapir young have characteristic white, lengthwise stripes. This apparently conspicuous pattern provides perfect camouflage in the animal's natural environment.

278

Brazilian Tapir (278)
Tapirus terrestris
Tapiridae

ranges from Colombia and Venezuela to Paraguay and to the Brazilian state of Rio Grande do Sul. The body is a uniform chestnut-brown or dark brown; a short mane covers the neck. The body is 190—200 cm long, the tail 8—10 cm; shoulder height is 85—95 cm; the weight reaches 200—250 kg. The Central American Tapir (*Tapirus bairdii*) is very similar to this species: it is the largest of all American species, reaching a length of 240 cm, a height of 120 cm, and weight of as much as 300 kg.

279

280

Mountain Tapir (279) is the smallest existing tapir, 160–180 cm long, 80–90 cm tall, and weighing 160–220 kg. It inhabits mountainous areas of Colombia, Ecuador, northern Peru, and western Venezuela to altitudes of 4,500

Tapirus pinchaque
Tapiridae

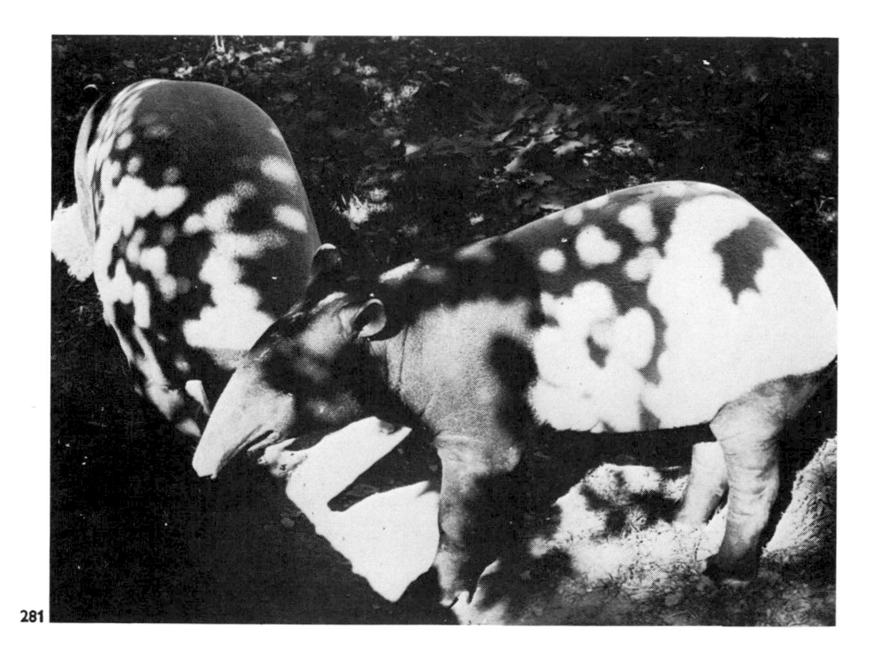

281

282

metres, that is to the upper tree line. Its body is brown above and pure white under the head and on the throat and chest. Unlike the rough, short and flattened hair of other tapirs, its hair is longer, softer and wavy.

Malayan Tapir (280, 281) is the largest species of the family. It reaches 260 cm in length, 130 cm in height, and 320 kg in weight. It is found in Burma, south-eastern Asia, in the Malayan Peninsula, and Sumatra. The front of the body and limbs are dark-grey to black, the rest is white or whitish-grey. This is protective colouration, breaking up the outline of the body and advantageous in the shade of forests (281).

Tapirus indicus
Tapiridae

Przewalski's Horse (282, 283) is a typical representative of the last family of perissodactyls, the horses. All horses have a single toe, originally the middle one, which is a product of their long evolution since the early Tertiary Era. They have developed towards adaptation for life in open steppes, and for fast running. All are herbivorous, and stay in small groups or large herds. They usually give birth to a single calf; twins are an exception. Przewalski's Horse is the only genuine wild horse which has survived to the second half of the 20th century. It is native to the steppes of central Asia and Mongolia. The last remaining specimens, numbering several dozens of animals, may survive in steppes and semi-deserts in south-western

Equus przewalskii
Equidae

Mongolia. On the other hand, this species does well in zoological gardens, where it lives longer than in the wild. Basic colouration of the body is ochre-brown to dark brownish ochre; the underparts are slightly paler, and the area around the muzzle is white. The back is divided by a narrow, almost black stripe and the limbs are sometimes indistinctly marked with dark stripes. The mane is erect, but relatively long. The coat is short and close-packed in summer and considerably longer, softer and wavy in winter. Body length varies between 220 and 260 cm; the tail is some 80—110 cm long; weight is 200—300 kg; and height is 130—140 cm. Przewalski's Horse lives in small groups of three to ten, twenty at most. Gestation lasts 328—343 days.

African Wild Ass (284)
Equus africanus
Equidae

was native to northern Africa, from the Atlas mountains to Egypt, the Sudan, Somalia, and Ethiopia. It has become extinct in most of these territories, and remains only in Somalia, and probably in the Danakil Desert in Ethiopia, in the form of the Somalian Ass (*Equus africanus somaliensis*) (284). The basic body colour is grey, and greyish white around the muzzle and on the underparts. The limbs are patterned with pronounced, dark stripes, and the back bears a narrow, lengthwise stripe. The mane is comparatively short. The tail is not entirely covered by the long 'horsehair' of the tails of horses: only the tip bears a tuft of long hair. The body measures 210—230 cm; the tail measures about 45 cm without the hair tuft and up to 75 cm with it; the height at the shoulder

283

284

is 130—135 cm; weight is 200—230 kg. Gestation lasts 360—370 days.
The Wild Ass lives in groups of ten to fifteen; these settle to graze in semi-desert locations.

Asiatic Wild Ass (285) was once widespread in Mongolia, western China, central Asia, Iran, south-western India, and Syria but it has become extinct in many places. The Central Asiatic Wild Ass or Kulan (*Equus hemionus kulan*) (285),

Equus hemionus
Equidae

285

286

The Quagga, exterminated in the second half of the last century, differed from other zebras by darker colouration and stripes only in the front part of the body

living in central Asia, is a well-known form and the least endangered. It is tawny-coloured, often with reddish tones, and paler below. Its mane is short, brown, continuing along the back in a broad, dark brown stripe. The body is about 210–220 cm long, the tail measures 45–48 cm, shoulder height may reach 130 cm, and the weight is 160–180 kg. Gestation lasts 325–335 days.

287

288

Kiang (286) is the largest of all Asian and African wild asses. It is 220—250 cm long; the tail without its horsehair measures 50—53 cm; the height is 130—142 cm and the weight is 200—280 kg. The basic colouration is chestnut to greyish brown; the underparts are strikingly pale, almost white. The mane is short and almost black, like the stripe along the centre of the back. Dark horsehair grows on the tip of the tail, and near its root. The Kiang inhabits mountain plateaux of Tibet, from the northern Himalayan slopes to the southern borders of the Gobi Desert. Gestation lasts 350—370 days.

Equus kiang
Equidae

Grevy's Zebra (287) belongs to the group characteristic of the African continent. It is the largest zebra, reaching 160 cm in height, and weighing up to 350—400 kg. It lives in groups of five to fifteen individuals in bush areas of southern Ethiopia, Somalia, the southern Sudan, and northern Kenya. Its stripes are very regular, black, narrow, and densely distributed, extending down the limbs to the hoofs; they are lacking on the belly and around the root of the tail. The ears are large and wide, endowing Grevy's Zebra with a distinctive appearance. Gestation lasts 380—390 days.

Equus grevyi
Equidae

Mountain Zebra (288) is a close relative of the preceding species, although it lives at the opposite end of Africa. It has the most southern distribution of all zebras. At present, it occurs very rarely in the mountains of Cape Province, and in the western coast of southern Africa. It is the smallest of the zebras, not exceeding 125 cm in height and 150—200 kg in weight. Groups of five to twelve animals sometimes gather into herds of 50 head. Stripes on the body are narrow, not very regular, except on the thighs, where they are distinctly wider. There is a characteristic dewlap on the throat. The Mountain Zebra occurs in two geographical forms: Cape Mountain Zebra (*Equus zebra zebra*), of which only a few hundred are left in the wild, and the more abundant Hartmann's Mountain Zebra (*Equus zebra hartmannae*) (288).

Equus zebra
Equidae

289

290

291

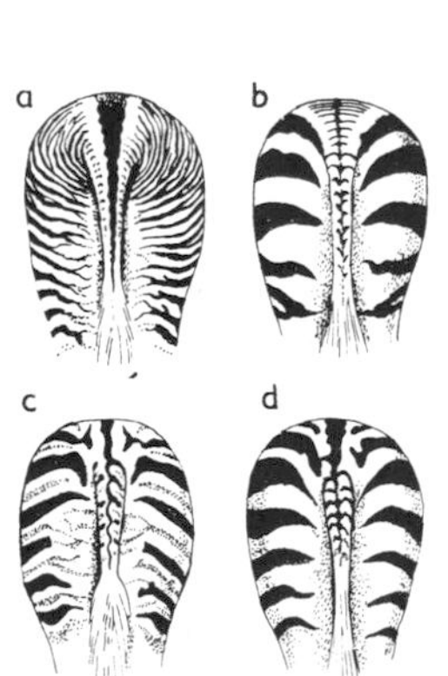

Stripes of various forms of zebra (view of the hind-quarters):
a) Grevy's Zebra,
b) Mountain Zebra,
c) Chapman's Zebra,
d) Böhm's Zebra

Steppe Zebra (289, 290, 291) is the most widespread and best known zebra. It is found in several geographical forms in steppes of southern, south-western and eastern Africa northwards to the Sudan and southern Somalia. In terms of size, the Steppe Zebra ranks between the two preceding species. The stripes are less numerous, generally wider and not always black, but also brown or dark grey-brown. In some forms, the stripes gradually fade out on the limbs and do not reach the hooves; the rump sometimes has less distinct brownish stripes between the regular black ones. The ground colouration may have a reddish shade. The most regularly striped forms include Böhm's Steppe Zebra (*Equus burchelli boehmi*) (291), widespread in eastern and south-eastern Africa. A less pronounced pattern of dark stripes occurs in Chapman's Steppe Zebra (*Equus burchelli antiquorum*) (289, 290) which ranges from south-western Africa and southern Angola, across Bechuanaland to Transvaal. The last species of zebra, exterminated in the second half of the last century, was the Quagga (*Equus quagga*), which used to live in steppes of the central Cape Province. It was striped only on the front part of the body, while the hind part was uniformly pale or dark brown.

Equus burchelli
Equidae

292

Chapter 11 SEA COWS

Odysseus was well aware of snares set for mariners by Sirens, bewitching sea nymphs. By their sweet singing, they lured sailors to their island; if the vessel followed the alluring, treacherous voices, it was wrecked on sharp coastal reefs concealed underwater. When the Sirens' island appeared on the horizon, Odysseus stopped the ears of all his men with wax to make them deaf to the enchanting singing. He ordered the men to tie him fast to the ship's mast. When he heard the captivating voices, ropes prevented him from throwing himself to the sea and swimming to the island of the perfidious nymphs. This is Homer's depiction of an adventure of the classical hero Odysseus: Sirens are described as beautiful sea nymphs. However, if Sirens, or rather sirenians, are discussed by a zoologist, their characteristics will differ totally from those presented by the classic of Greek Mythology, and the reader or listener will recall another name, which this group of mammals has received: sea cows.

Sirenians (Sirenia) are an ancient order of aquatic mammals, of which only a few representatives have survived. The whole order includes only five recent species: one of them has been totally exterminated by Man. Like the cetaceans, sirenians are restricted exclusively to an aquatic environment. Like cetaceans, sirenians have a spindle-shaped body terminated by a broad, horizontally flattened tail fin, although the hydrodynamic properties of the body lack the perfection attained by cetaceans. The skeleton of sirenians bears many signs proving a closer kinship to land mammals than cetaceans could claim: for example, the forelimbs have been transformed into fin-like organs, but all five toes have remained in the skeleton and have retained all their phalanges: occasionally, even stunted nails can be discovered in the skin. The shoulder section of the forelimb has also been well preserved; the elbow joint is fully developed, and the limb can be bent or stretched — unlike in cetaceans. The forelimbs are not used merely as fins, but also for digging at the sea bottom and tearing off seaweeds from the sea bed and rocks. All sirenians are harmless herbivores. They live in larger or smaller groups and feed by grazing water vegetation, mainly on the bottom of the sea. Their specialization to a vegetarian diet led to the development of a composite stomach similar to that of bovines. The teeth are often greatly reduced, and the food is ground instead by hard, horny plates located on the palate, the front part of the lower jaw, and on the tongue. In males of the Dugong, the upper incisors are transformed into short tusks, unseen when the jaws are closed together. Although the skin of sirenians appears to be almost bald, the reduction of the body hair cover is less advanced than in cetaceans. Moreover, sirenians grow rather tough and thick tactile hairs on their upper lips; these hairs are highly mobile and can assist the transfer of food into the mouth. These interesting aquatic mammals live predominantly in coastal waters from which they often penetrate into river deltas, and sometimes, swimming against the flow, far inland. In floods, some species even reach coastal swamps and lakes. They regularly give birth to a single young, born in the water. The legend of the Sirens was probably started by the fact that, unlike the majority of other mammals, females of these sea mammals have only one pair of mammary glands, located on the chest, and thus resemble the female human form. It remains only to be added — however odd it may sound — that, evolutionarily, the closest relatives of sirenians are elephants.

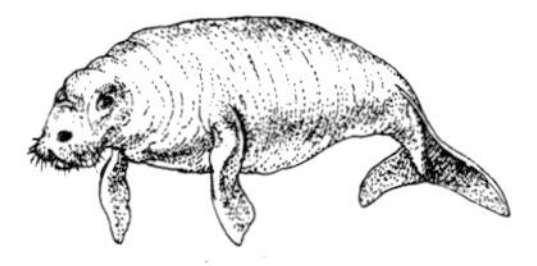

Steller's Sea Cow was exterminated by the second half of the 18th century – only 27 years after its discovery

Indian Dugong (292) is the only member of the first sirenian family (there are three of them).
Dugong dugon
Dugongidae

Dugongs are distinguished by a wide, deeply forked tail fin. The skin is brownish or greyish and it is only very sparsely covered with individual hairs, making it appear to be completely bare. The Dugong lives in coastal waters of the Red Sea, off the shores of eastern Africa, Madagascar, the Bay of Bengal, the Malayan Archipelago, the Moluccas, the Philippines, New Guinea, and northern Australia. It was, however, completely exterminated in many of its original habitats and its occurrence is very scarce even in places where it still does live. Dugongs form small groups, consisting mostly of two to five individuals, grazing the growths of sea plants and algae. They often tear whole clumps of vegetation from the sea floor, shaking them vigorously until the remains of sand and mud are removed. The Dugong is a large mammal, reaching 280—320 cm in length, and weighing as much as 200—240 kg. Gestation is estimated at 330—350 days.

The only member of the second family, Hydrodamalidae, was Steller's Sea Cow (*Hydrodamalis gigas*). It was also the only sirenian known to have lived in cold waters. Like the Dugong, it had a deeply forked tail. This sea cow is, unfortunately, that species, figured on p. 243, which was completely exterminated by Man in the not-too-recent past. It was discovered in 1741 by Captain Vitus Bering on the Commander Islands in the Bering Sea, but, by 1768, even before it received its scientific name and description in 1780, it had been entirely wiped out. It thus took man only 27 years to exterminate one animal species! Although the original number of Steller's Sea Cows probably did not surpass 5,000 to 5,500, Man's guilt cannot be excused. Mariners mercilessly killed the huge, defenceless animals, more than 750 cm long and weighing more than 4.5 tons, mainly as a source of palatable flesh and fat. According to old observations, especially to those of the naturalist Steller, accompanying Bering in his expeditions, the period of gestation lasted more than 12 months. It is interesting to note that both the extinct Steller's Sea Cow and members of the next family, the manatees, have only six cervical vertebrae, although, except for Hoffman's Sloth (*Choloepus hoffmanni*) which also has six cervical vertebrae and the Three-toed Sloth (*Bradypus didactylus*) with eight or nine cervical vertebrae, all other mammals have the identical number of seven cervical vertebrae.

Wide-nosed Manatee (293) is a representative of the last sirenian family, that of manatees. It
Trichechus manatus
Trichechidae

exists in three species, widespread in both western and eastern hemispheres. The Wide-nosed Manatee lives in the western hemisphere, ranging from the south-eastern coasts of the United States to the north-eastern shores of South America, and to islands in the West Indies. It is usually coloured a dingy grey to black. As in all other sirenians, the body seems to be completely bare, but it is in fact covered with sparse hairs. The hairy cover is more dense in this species than in the Dugong, the individual hairs being approximately 2 cm apart. These manatees usually live in groups comprising no more than 20 animals. However, groups of as many as 35 individuals have also been reported. Body length reaches 350—400 cm; weight can surpass 500 kg, exceptionally as much as 600 kg. Most specimens, however, are no more than 300 cm long, and weigh 200—360 kg. Unlike that of the Dugong and Steller's Sea Cow, this manatee's tail is not bifurcated, but has a rounded border, and it is less wide. Gestation in manatees, according to not very reliable sources, lasts 150—180 days; more plausible data suggest 355—365 days.

In concluding this chapter, it should be mentioned that this group of mammals had reached an evolutionary dead-end, and would probably have been in decline anyway. Nevertheless, it is a pity that man accelerated so drastically the end of these interesting creatures.

293

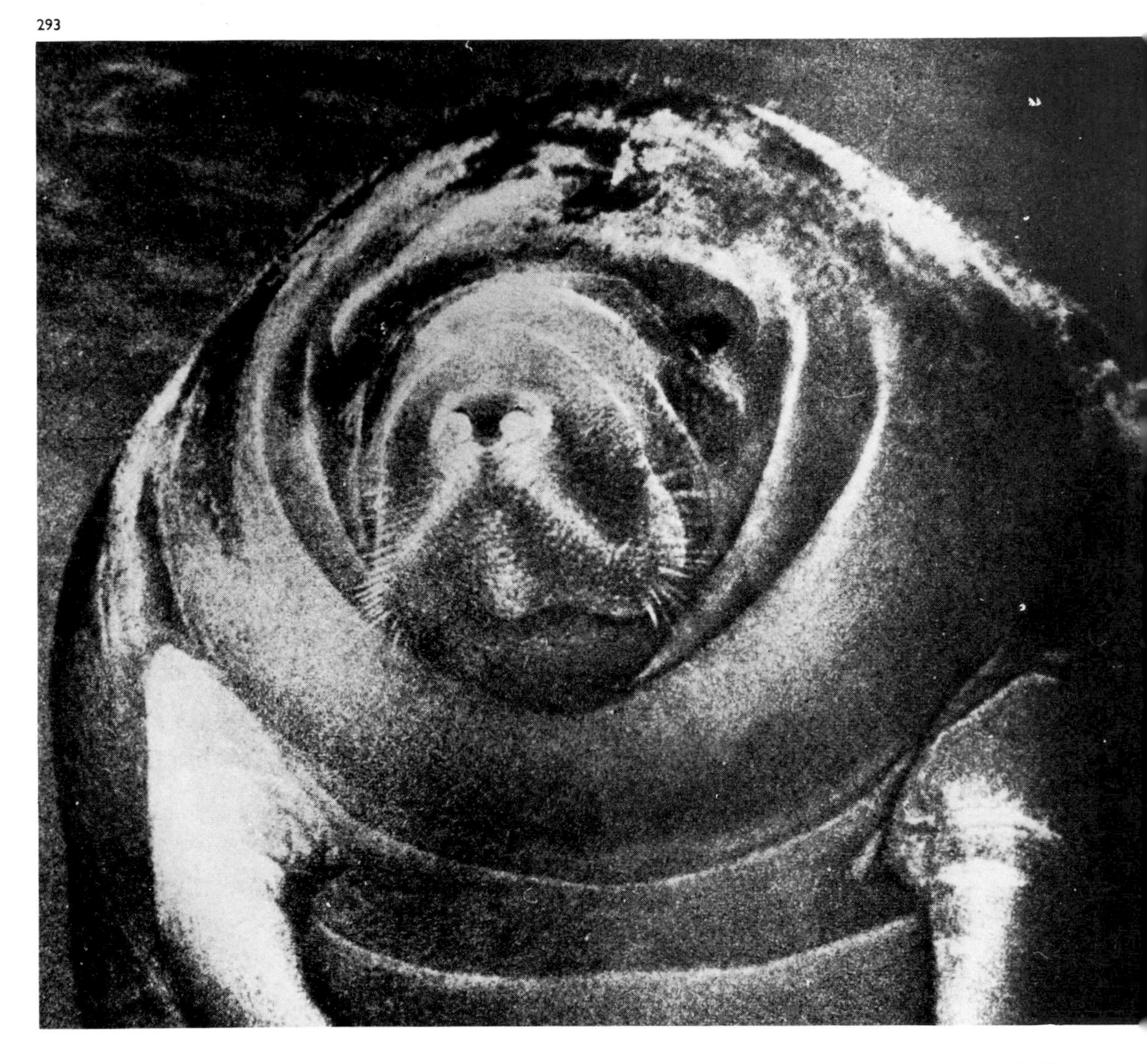

294

Chapter 12 HYRAXES — A ZOOLOGICAL PUZZLE

Like the sirenians, the following mammal order — hyraxes (Hyracoidea) — are a very ancient group in evolutionary terms. The order differentiated in the early Tertiary, and passed its peak a long time ago. The whole order is nowadays represented by the single hyrax family, Procaviidae, incorporating seven modern species. The hyraxes, also called 'dassies', are tailless, hare-sized herbivores: it is therefore all the more surprising that their skeletal structure and internal organs indicate certain evolutionary relations to perissodactyls and to elephants. For instance, the upper incisors are modified into short, downward-turned tusks, protruding from the mouth of the male. No wonder, then, that the systematic and evolutionary relations of this mammalian group used to mystify zoologists: hyraxes had even been ranked among rodents. They received the status of an independent order only in the first half of this century. As in perissodactyls, the greater proportion of the body weight is carried by the original third toe of each limb. It should be noted that the original five digits have been reduced: forelimbs have four toes, hind limbs only three. The last digital segments are equipped with small hoofs, but the middle toe of the hind feet is tipped with a claw, split lengthwise and serving for cleaning the fur. Hyraxes have two special features, lacking in other mammals. They have three vermiform appendices, and eyes endowed with a nictitating membrane, that is with a third lid, which closes across the eye from the inner corner. Hyraxes have relatively short but thick hair, predominantly grey, brown-grey or brown-coloured. Long tactile hairs grow above the eyes, on the sides of the head, on the nose and chin. Further tactile hairs can be found in the fur on the flanks and back. Hyraxes are good climbers and the tactile hairs are highly useful since they help them to recognize the nature of the environment and enable them to move with confidence.

295

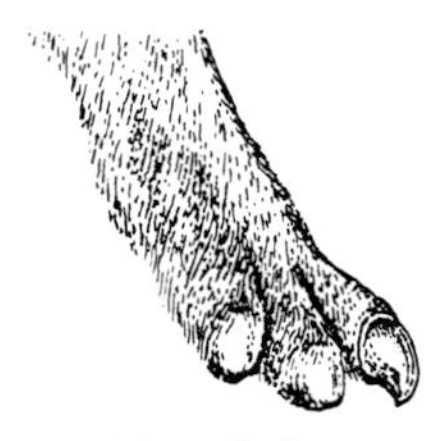

Hind foot of a hyrax

Rock Hyrax (294) inhabits all southern Africa, reaching north as far as Rhodesia, Angola, and southern and central regions of south-western Africa. The length of the stout body varies between 45 and 50 cm, the height at the shoulder does not exceed 26—28 cm, and the weight is about 3—4 kg. Colouration is brown; approximately in the middle of the back is a dark brown to black spot, surrounding the dorsal scent gland which is present in all hyraxes. The diet of hyraxes consists of fruits, leaves, bark of trees and bushes, etc. The Rock Hyrax, like the other species, is an excellent climber, both on trees and rocks. Climbing is facilitated by a 'climbing device' on the underside of the limbs; the soles are covered with elastic pads having numerous lengthwise and crosswise ridges; if a hyrax presses its limb to a surface, air is partly expelled from between the ridges, and the sole adheres to the surface, like a suction pad, sufficiently well to allow the animal to climb very steep rocky walls, stones, or boughs of trees. Hyraxes are found even high in the mountains, for example in Kenya, where another species — Johnston's Hyrax (*Procavia johnstoni*) — has been observed on Mount Kenya at an altitude of 4,500 metres. The Rock Hyrax frequents mostly dry, bushy territories, stony and rocky regions, but sometimes also the edges of thick forests. It lives in groups of several dozens. Duration of gestation is surprisingly long for an animal of such a relatively small size: 220—230 days; the litter averages three or four, exceptionally five young.

Procavia capensis
Procaviidae

Yellow-spotted Hyrax (295) is widespread on the Sinai Peninsula, in a large part of Africa, from Egypt and the Sudan across all eastern Africa to Zambia, Mozambique, eastern Angola, Transvaal, and north-eastern Bechuanaland. Unlike the preceding species, it has grey ground colouration, and the dorsal spot is bright yellow to ochre, or even whitish yellow. The length of the body is 45—52 cm, height at the shoulder is 28—32 cm and the weight varies between 3.5 and 4.5 kg. The Yellow-spotted Hyrax is mainly resident in arid savannas, seeking, in particular, sites with heaps of stones and boulders, solitary rocks, termitaries, or tall solitary trees. In western parts of eastern Africa, it is found at altitudes over 2,000 metres above sea level. It lives in groups. The diet is similar to that of the preceding species, gestation likewise. Litters number two or three young, exceptionally four.

Heterohyrax syriacus
Procaviidae

Tree Hyrax (296) is restricted by its way of life to densely wooded areas and tropical forests. Its range of distribution is vast: it covers most of Africa from the Cape Province to Kenya and Uganda in the east, to southern Cameroun, Nigeria, Ghana, and the Ivory Coast, as far as Liberia and Guinea in the west. Its colouration is brown, reddish brown to black; the hair around the dorsal gland is long and white. The body measures 45—55 cm, the height at the shoulder is 30 cm, and the weight is 3.5—4.6 kg. The Tree Hyrax is probably the most agile member of its family. It feeds on leaves of trees and bushes, and on epiphytic ferns. Gestation lasts about 225 days; a single young is produced, rarely two.

Dendrohyrax dorsalis
Procaviidae

296

As mentioned above, hyraxes form a small but highly specialized group of mammals characterized by certain evolutionary relationships with perissodactyls on one hand and with proboscideans on the other. This is also borne out by the level of their intelligence. The early naturalists thought that hyraxes were rather dull animals, but in more recent years, since these interesting creatures began to be kept in captivity, it has become apparent that the degree of their intelligence is far from low and corresponds to that of some perissodactyls.

Thus, hyraxes represent a group which is remarkable not only for the appearance of its members and their body structure, but also because of their way of life and the level of their intelligence. They furnish us with a good example of the errors a zoologist could commit were he to postulate a mutual evolutionary relationship between animal groups, based simply on their outward appearance. Only a thorough scrutiny of all the various criteria by which animals can be classified as allies, can result in the recognition and explanation of relationships among the individual forms and in the understanding of the complicated ways in which the evolution of living organisms has developed.

297

Chapter 13 ELEPHANTS

There are few mammalian groups to rival the attraction of the 'terrestrial giants' — the elephants of the order of proboscidians (Proboscidea). With the aardvarks (Tubulidentata), they rank among the least numerous of mammalian orders: proboscidians comprise only two species. Yet, in the early Quaternary, their order was much better represented, and it reached its evolutionary peak in the late Tertiary: a wide range of different forms was distributed not only in the Old World (which is now the last refuge of modern species), but also in the New World. Only a few of the fossil species exceeded the size of the elephants of today, although the contrary is believed to be true. For instance, most mammoths, living in Europe during the last Glacial Epoch, never reached the dimensions of large specimens of contemporary elephants. The same is true of the popular Siberian mammoths.

The proboscis or trunk is the most conspicuous organ of an elephant. It compensates for the animal's short neck and serves not only for breathing, but also for grasping food, drinking, and as a defence weapon. The teeth are no less remarkable: the tusks are naturally the most prominent ones, being in fact modified incisors which sometimes reach considerable length. They help the elephant in foraging for food: in digging out roots, crushing tree trunks, peeling bark, etc. As weapons, they are used less often than it is thought. When duelling, males fight mainly with the trunk and body. In addition to incisors (the tusks), elephants have only molars — canines and premolars are absent. However, the molars have truly elephantine dimensions — they are 10 cm wide and up to 25 cm long. In terms of size, they can be compared to a house brick. Their occluding surface is ridged with numerous crosswise ridges. It is said that only one molar in each half of each jaw is functional. This is not quite true: at any time throughout their long life, elephants are using two molars in each half of each jaw. As one molar becomes worn out, it is gradually replaced by the next molar, growing from the back forwards, and pushing out its predecessor, which is lost successively, one ridge after another. This replacement of molars takes place five times in a lifetime. The first molar is usually detached at the age of 3—4 years, the second, growing with the first, is pushed out at 6—7 years. The third begins to grow in the 3rd year, and disappears in the 12th—14th years; the fourth molar appears between the 6th and 7th years and is lost in the 25th or 26th year; the fifth molar becomes active in about the 16th year of life, and the last remains disappear at the age of 42—44. The last, sixth molar develops around the 33rd year, and at 65, possibly a little later, the tooth is completely worn out. Elephants are exclusively herbivorous, feeding on green and juicy parts of various plants. They also eat fruits, tender tree bark, young twigs and shoots, etc.

Elephants are digitigrade, walking on the tips of hoofed toes. The soles of their feet are equipped with thick ligamentous pads, which make the elephants' steps much softer and more silent than would be expected for their size. Their hide is extremely thick and almost naked. The brain weighs 5,500 grams and it is highly convoluted — which reveals that elephants are remarkably intelligent animals; their general reputation for cleverness is not unfounded. They are also noted for their exceptionally long life-spans, although stories of this kind tend to be exaggerated; elephants generally do not survive after 60—65 years. Specimens older than 70 years are extremely rare. Elephants live in large or small herds, usually led by an old female. Old males are often solitary.

African Elephant (297, 298)
Loxodonta africana
Elephantidae

is the larger of the two modern species included in the single elephant family. Body length, including the trunk and 129—140-cm tail, reaches 700—800 cm; the height is 300—350 cm and the weight 5,000—6,000 kg. Females are smaller by one-fifth or one-quarter. The front feet have four hoofs and hind limbs have three in African Elephants; only some of the so-called Forest Elephants have five hoofs on front limbs and four on the hind ones. The trunk is terminated by two finger-like protuberances. Tusks are well-developed in both sexes, although they are regularly longer in males — 180—250 cm long in old bulls — approximately one-

298

third being hidden inside the hide and skull. According to the latest official data, the record-making pair of tusks was that of an Elephant killed in Kenya several decades ago. The tusks had the following dimensions: length of the outer arch 311.15 cm and 319.2 cm; maximum circumference 61.6 and 59.7 cm; weight 102.3 and 95.95 kg. Verified data state that the longest known tusk, also from Kenya, was 338 cm long, but its maximum circumference was only 49.5 cm, and weight 54.36 kg. The African Elephant was originally widespread in steppes, bush and forests of all Africa south of the Sahara, and the Libyan-Egyptian deserts and semi-deserts. However, it has been exterminated in a vast part of this territory. It has been protected over the last decades, but it is sad to state that in the first half of this century, Man killed half a million elephants in Africa, which represents 10,000 animals in one year! No further commentary is needed. Despite its colossal size, the African Elephant (and the Indian Elephant also) are far from being as clumsy as they might seem. It is true that they do not move faster than a speed of 4 to 6 km per hour, but when attacking, they can charge at a speed of 30 to 35 km per hour. It is observed that fighting elephants are capable, if not exactly of nimbleness, then certainly of manoeuvres one would never expect in such gigantic bodies. Several geographical forms are distinguished — the described typical or nominate form *Loxodonta*

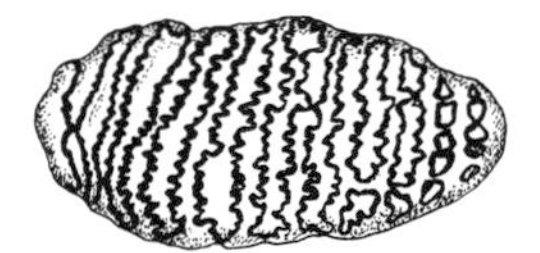

Lamellated molar of an elephant from above

africana africana is from southern Africa. Other well-known forms of the African Elephant are the Forest Elephant (*Loxodonta africana cyclotis*) (298 right), which lives in the virgin forests of central and western Africa, and the Bush Elephant (*Loxodonta africana oxyotis*) (297, 298 left) found in open and wooded savanna of eastern and south-eastern Africa. The Forest Elephant is noticeably small: the height of adult males varies between 220 and 240 cm, only rarely reaching 250 cm; the ears are smaller, and rounded at the lower end; the tusks are narrower than in Bush Elephants, and they are not arched upwards, but point downwards. The Bush Elephant, on the contrary, is the largest form, with large ears, elongated in the lower parts. Gestation of African Elephants lasts $21\frac{1}{2}-22$ months, that is 640–665 days; the female gives birth to a single young.

Indian Elephant (299, 300, 301) inhabits southern and south-eastern Asia. It is found only in a few regions of its ancestral home, which included southern, western and northern India, southern Nepal, Assam, Burma, south-eastern Asia, the southernmost China and Malayan Peninsula, Sumatra, and Borneo (Kalimantan). Indian Elephants are usually smaller than their African counterparts, differing from them in, among other things, a bulkier, but more proportionately built body and smaller and differently shaped ears.

Elephas maximus
Elephantidae

299

Teats in elephant females are not situated in the lower abdomen as in most mammals, but on the chest as in sirenians, bats and primates

They have five hoofs on the front limbs and four on the hind ones, and only one finger-like protuberance on the trunk. Tusks in females are much smaller than in males, in many cases being totally absent or invisible externally. Tusks in males are not very massive either, and sometimes do not develop at all. Adult males measure 650—800 cm including trunk and the tail which is 110—140 cm long; the height is 270—325 cm and the weight 3,500—5,000 kg, exceptionally more. When well-developed, the tusks of old males are 110—180 cm long; one-third is always concealed inside the hide and skull. The largest pair of tusks recorded measured 266.7 and 260.4 cm, the maximum circumference was 55.5 and 55.9 cm, and their weights were 72.98 and 72.52 kg. This pair was from an elephant shot by the English king, George V, in the western part of the Nepal-Indian border. From antiquity, Indian Elephants have been used for labour, and often exploited as awe-inspiring animals of war. In India and south-eastern Asia, elephants are still employed for carrying heavy burdens, dragging felled trees, and so on. Until recently, they were also used in hunting expeditions, particularly in tiger hunts. So-called 'white elephants' are often spoken of written of. True albinos, however, are not known to exist. The 'white elephants' are mostly buff-coloured or light-spotted individuals which are often regarded by native

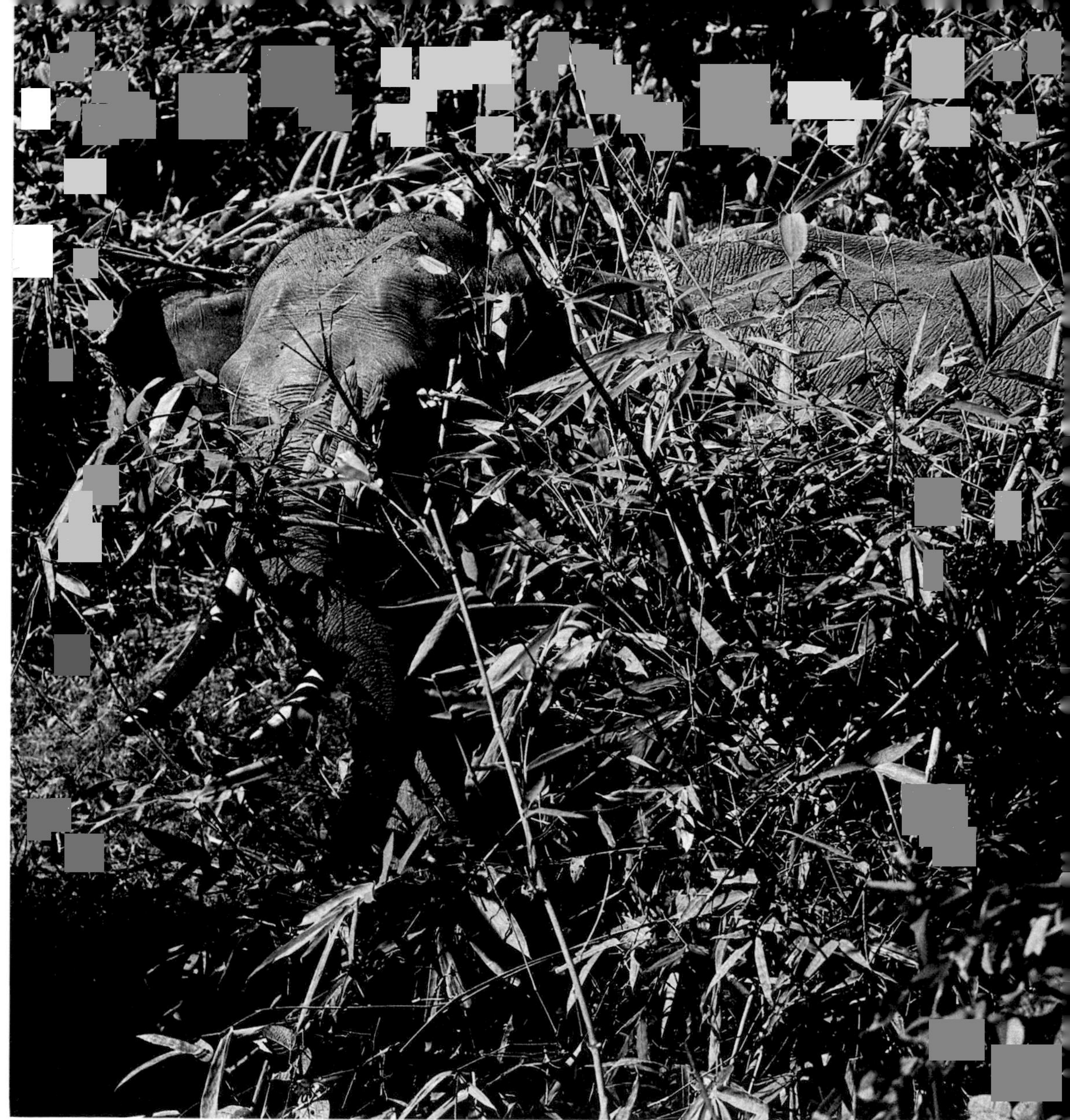

301

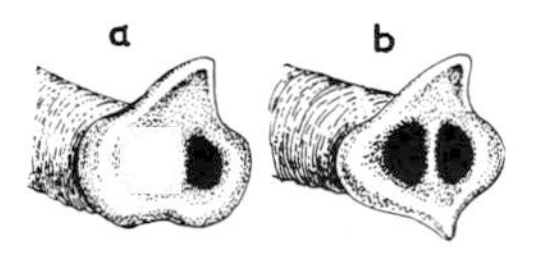

Tip of the trunk in Indian Elephant (a) and in African Elephant (b)

people as 'kings' among elephants or even sacred animals. According to the experience of Indian breeders, gestation in the Indian Elephant varies: young females are usually born after 610—620 days of pregnancy, young males after 645 to 670 days. A single young is usually born, but twins have also been reported.

302

Chapter 14 HARES AND THEIR ALLIES

Hares, rabbits, and piping hares used to be classified as rodents, but this has been refuted by research carried out in the past decades: the relationship between hares and rodents, presumed to be evolutionary, proved to be merely physical. They were then classified into the independent order Lagomorpha. Although both hares and rodents have corresponding features of adaptation, for example the first pair of chisel-like incisors grow throughout life, and they are separated from the molars by a large gap (the diastema), this feature has developed independently in the two groups, in relation to their dietary specialization. Moreover, hares have another pair of tiny, peg-like incisors in the upper jaw, and they differ from rodents in the structure of the molars. Palaeontologists discovered that the primary types of rodents and hares were differentiated to such an extent that they must have developed independently from the very beginning, probably from different ancestors. Finally, results of serological studies totally deny any relation of hares to rodents, suggesting rather their kinship with artiodactyls (even-toed ungulates).

Despite their ancient origin and a rather small number of species (only some 60 species), hares certainly are a successful group, which has settled over all the main continents except Australia.

American Pika (302) represents the more ancient family of piping hares, including small types,
Ochotona princeps
Ochotonidae

widespread as relicts only in North America and Asia. The Pika is one of the two species of American piping hares, and it is found in Canadian and US mountains. It weighs up to 500 grams.

Daurian Piping Hare (303) is one of the 16 species of piping hare inhabiting vast parts of Asia,
Ochotona daurica
Ochotonidae

from the easternmost edge of Europe in the Volga region to Japan, Tibet and Burma. They frequent stony slopes in mountains, forests, steppes, and semi-deserts, where they dig burrows. They are highly cold-resistant, being active in the snow at 17°C below zero, and only certain species pass a period of a light winter sleep. Food consists of green parts of plants, dried up in autumn and stored.

European Hare (304), like all hares, is well-adapted to fast running: it has long hind legs. It is
Lepus europaeus
Leporidae

widespread over almost all of Europe, in western Africa, and in western Asia. Its original territory is still enlarging in line with agricultural expansion; in addition, the European Hare was introduced to North and South America, Siberia, Australia and New Zealand. Except during the

303

Skull of a hare

304

mating season, it is solitary, hiding in shallow surface dens. The breeding season lasts from January to October; females produce one to five young in three or four litters a year. The young are born with open eyes, covered with hair, and they soon become independent. It weighs 3—5.5 kg.

Blue Hare (305) weighs 2—4.5 kg. It is dark brown with a bluish tone in summer, and snow-white in winter. It is found in forest areas of Eurasia, and as a relict in the Alps. It lives on green food and in winter on the bark of trees and bushes.
Lepus timidus
Leporidae

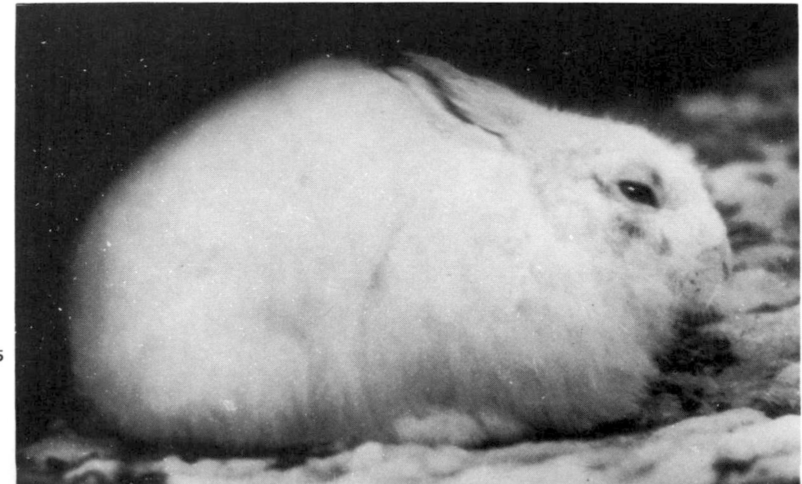

305

306

Snowshoe Hare (306) is a close relative of the Blue Hare, and also changes its summer colour to the winter white. It is resident in coniferous and mixed forests of North America. It overbreeds at intervals of 10—12 years.

Lepus americanus
Leporidae

European Wild Rabbit (307) is the wild ancestor of all breeds of domestic rabbit. Native to the western Mediterranean, north-western Africa and Spain, thanks to introduction it later spread over western and central Europe. Its excessive reproduction in Australia and New Zealand became a problem in agriculture. In Europe, it was recently almost eliminated by repeated epidemics of myxomatosis, a fatal viral disease. The rabbit inhabits dry, sandy sites on gentle slopes and fallow ground and is often found in city parks. It lives in colonies, digging deep burrows. Rabbits are extremely prolific, reach early sexual maturity, and have a short gestation (28—31 days). They bear four to seven litters yearly, dropping four to twelve young. It is much smaller than the European Hare, weighing only 1.5—2.5 kg.

Oryctolagus cuniculus
Leporidae

307

308

Chapter 15 ANTELOPES AND THEIR ALLIES

Most people take herbivores for harmless, gentle creatures. The following order of mammals – artiodactyls (Artiodactyla) – consists only of herbivores, and many of their species do not always show much docility. They defend their lives or those of their offspring so fiercely that even lions or tigers often show them respect. Old hunters were right in saying that an enraged buffalo is the most dangerous animal of all. Furthermore, herds of some hoofed mammals could literally turn green savannas into deserts, were it not for carnivores regulating their numbers.

While perissodactyls are characterized by the mesaxonic limb, the artiodactyl order is typified by the 'paraxonic' limb – the axis of the limb passes between the third and fourth digits, which are of equal strength, and the proportion of the body weight borne by the limb is divided evenly between the two digits. Of the other digits, the first toe or thumb is always lacking, the second and fifth digits are less developed or even stunted (except in the hippopotamus), and are shorter, so that the animal does not walk on them. In many cases hoofs of the second and fifth toes are completely absent and sometimes, as in camels, the toes themselves are totally reduced.

Almost all artiodactyls are exclusively herbivorous, as is evident from their teeth, and a very few forms can be described as omnivorous (pigs, peccaries). Artiodactyls are classified into three suborders: non-ruminants (Nonruminantia), including three families; camels (Tylopoda) with one family; and ruminants (Ruminantia), including five families. This abundance of families indicates that artiodactyls are a mammalian order with a rich variety of forms, differentiated in shape and anatomy, and all specialized for the herbivorous way of life which is most pronounced in ruminants. Their stomach is divided into four compartments: rumen, reticulum, omasum (also called 'manyplies' or 'psalterium'), and abomasum; each part has a characteristic lining. This arrangement and the fact that the partly digested cud is regurgitated from the front compartments of the complex stomach into the mouth and chewed again, facilitate efficient processing and digestion of food.

309

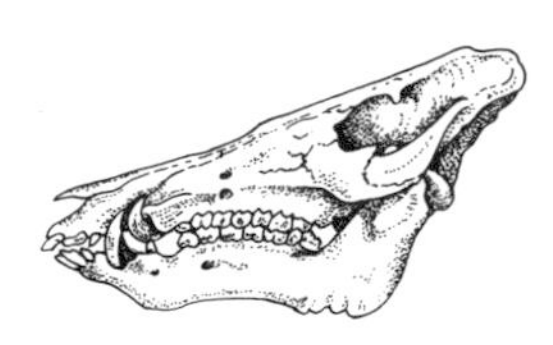

Skull of the Wild Boar

310

311

312

Wild Boar (308) is a typical representative of the suids, the first family of non-ruminants. It is a stout animal having many geographical forms which inhabit part of northern Africa and a large Eurasian territory from the shores of the Atlantic to the Far East of the former USSR and south-eastern Asia. Old males of the largest forms measure up to 185–220 cm (including the 25–30 cm tail), and weigh 200–250 kg, sometimes even more than 300 kg. Females are only two-thirds to three-quarters this size. Like other members of the family, the Wild Boar has highly developed upper and lower canine teeth, protruding from the mouth especially in males; the upper canines arch upwards. They are lethal weapons, fatal even for young, inexperienced tigers which may incautiously attack an old boar. The Wild Boar is omnivorous, foraging for roots, fruits, bulbs, mushrooms, insect larvae, worms, molluscs, various small vertebrates; it also eats dead fish and carcasses of larger animals. It frequents mostly forests, but can be found also in open country. Following 130–140 days of gestation, the female gives birth to five or six, but sometimes as many as 14 young.

Sus scrofa
Suidae

Bush Pig (309, 310) is restricted to Africa south of the Sahara, occurring in a number of different geographical forms such as the West African Bush Pig (*Potamochoerus porcus porcus*) (309), or the South African Bush Pig (*Potamochoerus porcus koiropotamus*) (310). The Bush Pig is smaller than the Wild Boar, rarely weighing more than 120–130 kg. Gestation lasts 115–130 days; the litter consists usually of two to four, sometimes as many as eight, young.

Potamochoerus porcus
Suidae

Wart Hog (311) has a characteristically shaped head with several outgrowths in the shape of huge warts, and with two pairs of powerful and long canine teeth. The Wart Hog lives in Africa south of the Sahara, except for tropical primary forests and southern and south-western parts of southern Africa. Its total length is 125–140 cm; its weight is 75–100 kg. Gestation lasts 170–176 days and three or four young are born.

Phacochoerus aethiopicus
Suidae

Giant Forest Hog (312) reaches almost the same dimensions as the largest forms of the Wild Boar. It has a large, wide head with massive, but not very long tusks. The hair is black-brown to black, coarse and sparse. This comparatively rare suid species lives in forests and jungles of central tropical Africa.

Hylochoerus meinertzhageni
Suidae

313

The Giant Forest Hog lives in small family groups; plant food dominates their diet. Gestation lasts 125 – 130 days; the litter numbers two to six young.

314

315

Babirussa (313) is confined to northern Celebes, the adjacent islands of Togian and Sula, and to Buru Island. This almost naked brownish suid reaches a total length of up to 150 cm, including the tail, which is 30–35 cm long. It weighs 100 kg at the most. Unlike other suids, its upper canines (tusks) do not grow up from the sides of the upper jaw, but grow through it and protrude from the upper surface of the snout, usually hooking backwards. In old males these tusks can reach up to 30 cm (measured along their outer arches). The lower tusks are also of considerable length. In females both upper and lower tusks are substantially shorter. The arching of the upper tusks gave rise to a native legend of babirussas hanging by their tusks in tree branches at night. Rather an amusing idea, but entirely false. The Babirussa is restricted to rain forests. It lives in small family groups, feeding on various roots, fruits, on insect larvae and small animals. Gestation lasts 147–152 days; one or two young are born.

Babyrousa babyrussa
Suidae

Collared Peccary (314) represents the peccary family, which includes only three species. One of them (*Catagonus wagneri*) was surprisingly discovered in Paraguay, in the Gran Chaco region, in the first half of 1970's. Peccaries differ from suids in having a very short, stumpy tail, and upper canines which point downwards as in carnivores. The Collared Peccary is distributed from the south-east corner of the United States across Central America to central Argentina. Including the 3–5 cm tail, its body measures 90–100 cm; it weighs 16–28 kg. Peccaries live in groups of 25 to 30, probably up to 50 individuals. Gestation is 142–149 days; twins, exceptionally three young are born.

Tayassu tajacu
Tayassuidae

Pygmy Hippopotamus (315) belongs to the last family of non-ruminants, comprising only two species. Hippopotamuses have almost naked, slate-grey to brown skin, a large, angular head, and four fully developed toes on all limbs. The

Choeropsis liberiensis
Hippopotamidae

316

upper and lower canines grow into massive tusks, invisible when the mouth is closed. Hippopotamuses are herbivorous, adapted to a more or less amphibious way of life. The Pygmy Hippopotamus inhabits swampy forests of western Africa, mainly Liberia, Sierra Leone and Guinea, but it has become relatively rare. It lives singly or in pairs, feeding on aquatic and land vegetation, roots, young twigs, etc. It is less dependent on water than its larger relative, but despite that, whenever in danger it seeks refuge in water or swamps, not the forest. Body length is about 150 cm, the tail is 15—20 cm long, shoulder height is 75—83 cm,

and the weight is 170–250 kg. Gestation lasts 190–210 days; a single young is produced, always on dry land.

Hippopotamus (316) is substantially larger than the preceding species. It reaches a length of 420–490 cm, including the 50–65 cm tail. Weight varies from 2,500 to 3,300 kg. Tusks with roots can surpass 60 cm in length and 3.5 kg in weight. The reddish colouration of the skin, often noted in hippopotamuses, is caused by excretions of the skin glands which contain red pigment. The Hippopotamus used to occur near all large African rivers south of

Hippopotamus amphibius
Hippopotamidae

317

the Sahara, but not in tropical forests and south-African deserts. In the past it was found even in Palestine. Nowadays, it is extinct in many areas. Hippopotamuses are highly dependent on water. They are good swimmers and divers; they rest in water and obtain some of their food there but their main food, obtained at night, is land grasses and bushes. They live mainly in small groups, but sometimes form communities of several dozens of individuals. Gestation lasts 227–240 days; females give birth to one, exceptionally two young, which are born and suckled in the water.

Bactrian Camel (317)
Camelus bactrianus
Camelidae

represents the suborder of camels (Tylopoda), and its only family. Camels are evolutionally a relatively ancient group, originating on the American continent in the early Tertiary. Only two digits are developed (third and fourth), with a comparatively large walking surface, which prevents them from sinking into sand or other loose material. They have long necks and limbs, and they run with great stamina. Recent forms of camels inhabit Africa, Asia, and South America; all of them are adapted to life in arid regions of a semi-desert or mountainous nature, content with grazing on any kind of vegetation. The hump of true camels contains fat, and its size alters with the animal's physical condition. Camels can remain without water for a long time. Until recently, it was believed that when the camel is thirsty, its body chemistry releases water from the hump fat, but the mystery of the camel's resistance to thirst is based on its body temperature, which can vary from 34° to 40.5°C. The animal thus will tolerate a higher body temperature than other mammals before beginning to lose water by sweating. It has also fewer sweat glands than other mammals. The Bactrian Camel has been bred by man for many centuries;

318

it is a domesticated descendant of the Wild Camel (*Camelus ferus*). The domesticated Bactrian Camel lives mostly with the inhabitants of inland desert regions of the Asian continent (where its wild ancestor used to occur). It is 350—380 cm long, including the tail, which has a length of 55—65 cm. Height at the shoulder is 180—195 cm, and the weight is 450—690 kg. Gestation lasts 380—395 days. Usually a single young is born, rarely two.

Arabian Camel or **Dromedary** (318) is known only in its domesticated form; its wild ancestors are unknown. An earlier hypothesis that the one-humped Dromedary was bred from the two-humped Bactrian Camel did not prove to be true,

Camelus dromedarius
Camelidae

319

320

as can be appreciated in the fact that hybrids of the two species are either sterile or produce weak offspring, dying usually soon after birth. The Arabian Camel is kept mainly in northern parts of Africa and Arabia, but it is equally common in central Asia and north-eastern India. It was successfully introduced to southern Africa and also to Australia, where it acclimatized without any problem. The Arabian Camel is more slender and often taller than the Bactrian Camel. Total body length is 350 to 410 cm, including 60 – 75 cm of the tail; shoulder height is 180 – 225 cm; weight is 400 – 600 kg. The Arabian Camel is more drought-resistant: in hot weather it can live up to 6 days without water, while the Bactrian Camel survives only about 3 days. Gestation lasts 365 – 440 days; the female gives birth to one, rarely two, young.

321

322

Guanaco (319) represents South American camels or llamas. All llamas are considerably smaller than camels, they lack fat humps, and the walking surface of the two digits is narrower. The Guanaco originally lived in mountainous regions of the Andes (Cordilleras) from Tierra del Fuego to southern Ecuador, and in valleys and flatlands of northern and central Argentina. Nowadays, it does not reach farther north than southern Peru, and it has been exterminated in Argentina, except high mountain regions. The Guanaco,

Lama guanicoe
Camelidae

323

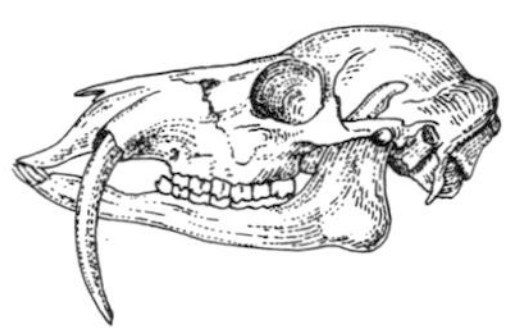

Skull of a male Musk Deer

324 325

like the following species, lives mostly in smaller groups of five to 15 females, led by a single male. Like camels, llamas are undemanding in terms of food; they find enough even in waste mountainous areas, at altitudes above 5,000 metres. The ground colouration is brownish to red-brown, whitish below. Total length is 175–225 cm, including the 25 to 30 cm tail, shoulder height is 90–115 cm, and the weight is 55–96 kg. Gestation lasts 320–335 days; usually a single young is born, rarely twins. Well-known domesticated forms of the Guanaco are the Domesticated Llama (*Lama glama*) and the Alpaca (*Lama pacos*).

Vicuña (320)
Lama vicugna
Camelidae

is the second wild llama species. It also inhabits high mountains of the South American Andes in Peru, western Bolivia, and northern Chile, where it can be found mostly at altitudes of 3,000–5,800 metres. In colour it resembles the Guanaco, but it is markedly paler. The chest and lower neck are covered by a mane of hair 20–35 cm long. Total body length is 165–200 cm, the tail is 15–25 cm long, height at the shoulder is 75–100 cm, and the weight is 35–65 kg. Gestation corresponds to that of the previous species; the litter almost always numbers a single young.

The first family of the suborder of ruminants (Ruminantia), the mouse deer family, is evolutionarily very ancient: it comprises four species of small ruminants living in central Africa (one species) and in southern and south-eastern Asia (three species). All mouse deer have well-developed second and fifth digits, slightly more slender and further from the ground than the third and fourth toes. Upper canines are modified into down-turned tusks with their tips projecting from the mouth. The hind limbs are substantially longer than the front ones, which makes the animal look somewhat hunched.

Indian Mouse Deer or **Chevrotain** (321) is a resident of central and southern India and Sri Lanka. It measures 48–60 cm, including a tail 3–5 cm long. Height at the shoulder is 25–30 cm, weight is 2.5–2.7 kg. Colouration is usually dark brown with light patches arranged in longitudinal stripes. It lives solitarily except during the breeding season. Gestation lasts 150–155 days; the litter contains one, less often two young.

Tragulus meminna
Tragulidae

Javan Mouse Deer (322) inhabits the southern part of south-eastern Asia, the Malayan Peninsula, Sumatra, Java, Borneo (Kalimantan), and some smaller adjacent islands. It is red-brown, somewhat smaller than the Indian Mouse Deer; it weighs usually 2–2.3 kg. Its habits, duration of gestation and number of young are the same as in the preceding species.

Tragulus javanicus
Tragulidae

326

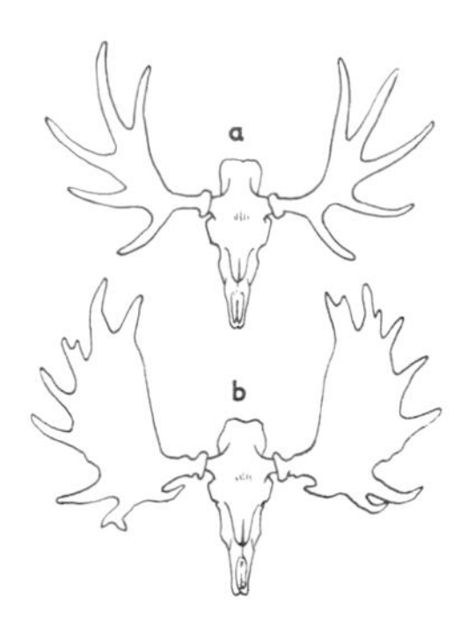

Two types of antler in the the Moose: tined (a) and palmate (b)

327

Chinese Water Deer (323) is one of the many members of the deer family, and one of the two
Hydropotes inermis species of the family which entirely lack antlers – the characteristic
Cervidae feature of males in all other deer. Antlers are growths of bony tissue and

328

329

are shed and renewed at more or less regular intervals of one year or more. The Chinese Water Deer, like the Musk Deer (*Moschus moschiferus*), has, instead of antlers, elongated upper canines, shaped like strong, sharp, curved tusks, protruding far out of the mouth. The tusks are present in both sexes, although they are larger and stronger in males. The coat is greyish to reddish brown, interwoven with fine, individual black hairs. The underparts are paler, almost white. The Chinese Water Deer lives in eastern and north-eastern China, mainly in the river basins of the Yangtse and Hwang-ho, and in Korea. It seeks places with tall grass, in bush and mountain valleys. The species can be encountered in the wild even in some regions of England and France; the animals are not, however, indigenous to these places, but are perfectly acclimatized descendants of animals escaped from captivity. The adult measures 80—105 cm including the 5—8 cm long tail, height at the shoulder is 45—55 cm, and the weight varies between 9 and 15 kg. Gestation lasts 180—200 days on average; usually twins are born, three young are rare, and four are extremely exceptional.

Muntjac (324), also called 'Barking Deer', is approximately the size of a Chinese Water Deer. It
Muntiacus muntjak
Cervidae

is native to southern and south-eastern Asia, from western India to the Sunda Islands and central and eastern China. The males have small, pointed antlers, and upper canines elongated into tusks, which also occur in females, but are considerably smaller. Muntjacs are solitary or live in pairs and feed on grass, leaves of bushes, and young shoots. They inhabit thickets and forests from lowlands to mountain valleys. Gestation lasts 175—182 days; a single young is born, rarely two.

330

Brocket Deer (325) inhabits dense tropical forests of most of Central and South America, approximately from the Tropic of Cancer to 30°S. Its coat is a dark red-brown. Antlers in males are short and unbranched. The body measures 95 to 130 cm, the tail is 10—15 cm long, shoulder height is 60—70 cm, and the weight is 16—21 kg. Gestation lasts most frequently 215—225 days; usually a single young is born; twins are rare.

Mazama americana
Cervidae

Mule Deer (326) occurs from western Canada and the United States southwards to northern Mexico. It seeks localities with thick vegetation. Its colouration changes slightly according to the season, ranging from tawny-grey to fawn. Its antlers are not very massive, but highly branched. The total length is 165—220 cm, including 17—30 cm of tail. Height at the shoulder is 100—115 cm; weight is 50—160 kg. Gestation generally lasts 196—210 days; most frequently twins, exceptionally three or even four young are born.

Odocoileus hemionus
Cervidae

Pudu (327) inhabits lowlands and the lower slopes of the Chilean Andes, and adjacent islands, occurring mainly in forests, between latitudes 50°S and 20°S. The Pudu's coat is grey, grey-brown to dark brown. Antlers are short and unbranched. The Pudu weighs 7—9 kg and measures 80—90 cm; the tail is 2.5—4 cm long, and the height at the shoulder is up to 40 cm. Gestation lasts 200—210 days on average; the litter regularly consists of a single young.

Pudu pudu
Cervidae

Moose (328) is the largest representative of the deer family. It is widespread in forest areas of North America, from Alaska and Canada to the north-western United States along the Rocky Mountains, and in Eurasia, where it occurs in Norway, Sweden, parts of central Europe, and east to Mongolia, Manchuria, and the Amur-Ussuri region. Old males reach 250—320 cm in length, the tail is 6—11 cm and the height at the shoulder is 155—190 cm; weight is up to 830 kg. The colouration varies from grey-brown to almost

Alces alces
Cervidae

331

black. The underparts and limbs are lighter-coloured, sometimes almost pure white. Throughout its vast area of distribution, the Moose has many geographical forms (subspecies), differing in colouration and size and in the massiveness of the antlers, which are usually typically palmately branched. The Alaska Moose (*Alces alces gigas*) (328) is one of the largest.

332

333

The Moose feeds mainly on parts of small trees and bushes, but also on bark, young shoots and aquatic plants found in water along the shores of rivers and lakes. It is solitary or lives in family groups. Gestation lasts 242–250 days; most often one or two young are born.

Reindeer (329) is a typical inhabitant of arctic and subarctic regions of both Old and New Worlds.
Rangifer tarandus
Cervidae

Large herds of wild reindeer, however, are now a thing of the past. Reindeer have been domesticated since long ago, and rank among the most important domestic animals of the Far North. Colouration of wild reindeer is mainly grey, tawny to brown, with lighter underparts. Body length varies from 130 to 200 cm, the length of the tail is 10–20 cm, height at

334

335

the shoulder is 110—125 cm, and the weight is 100—220 kg, exceptionally 300 kg. The Reindeer is the only deer with antlers in both sexes, although females' antlers are smaller. Wild reindeer used to live in large or small groups and herds. They were noted for their massive autumn migrations from open tundra to forest regions. In spring, they returned in smaller groups. Reindeer feed on tough tundra grass, leaves of stunted birches and willows, various lichens, mosses and arctic and subarctic vegetation. In the rutting season, reindeer males, like males of many other deer, fight duels and gather harems of five to 40 females. Gestation lasts 235—242 days on average; litter size is one or two young.

Père David's Deer (330)
Elaphurus davidianus
Cervidae

has become extinct in the wild, and survives only in zoological gardens and deer parks. The existing specimens are all descendants of a small group imported to Europe between 1869 and 1890 from the Chinese Imperial Summer Park of Nan Hai-tsu near Peking. By the last century, this park had already become the last and only refuge of Père David's Deer. During the so-called Boxer Rebellion in 1900, all deer in the park were killed. Their numbers in captivity increased in the first half of this century to such an extent that, in 1960, Père David's Deer was reintroduced to China from the English park at Woburn Abbey, after more than half a century. The history of this deer has proved to be a noteworthy example of human understanding and effort, which can help to maintain animal species for the generations to come. The original home of Père David's Deer is believed to have been the alluvial regions of north-eastern and central China. Its fossil remains were found also in Manchuria and southern Japan. The ground colouration of its coat is brownish grey, with a reddish overtone in summer. The form of its comparatively large antlers is unique among contemporary deer, testifying to a certain extent to the ancient origin of the species. The relatively long tail (45 to 55 cm) is also an outstanding feature. The body measures 140—170 cm, height at the shoulder is 100—120 cm, and the weight is 150—220 kg. Gestation lasts approximately 250—267 days; females give birth to one or two young.

Axis Deer or **Chital** (331) has its home in India and Sri Lanka. It has been introduced in New Zealand and in Australia, South America and elsewhere. Its favourite haunts are thickets near water. It lives either solitarily or, more often, in small or large herds (exceptionally, communities of 90 to 100 deer have been observed). Its body measures 120—150 cm, the tail 20—30 cm, height at the shoulder is 75—90 cm, and its weight is 70—90 kg. Females are pregnant for 210—230 days; they give birth to one to three young, most often twins.

Axis axis
Cervidae

Hog Deer (332) is related to the Axis Deer. It is widespread in northern part of India, in Assam, Burma, Thailand, Indo-China, and on the Calamian Islands in western Philippines. It was also introduced to Sri Lanka. The Hog Deer is somewhat smaller than the Axis Deer and it has comparatively shorter legs, which make it look more robust and stouter. It lives solitarily or in small herds, preferring thick, grassy jungle and forest margins with ample humidity. The female is pregnant for 220—235 days, delivering usually a single young.

Hyelaphus porcinus
Cervidae

336

337

Sambar (333) is resident in India and south-eastern Asia, southern and south-eastern China, on the Malayan Peninsula, on Sri Lanka, Sumatra, Borneo (Kalimantan), Hainan and Taiwan. Ground colouration of its coat is reddish or greyish brown to dark brown or almost black. A tough mane of longer hair grows on the throat. Body length varies in individual geographical forms from

Rusa unicolor
Cervidae

338

339

The antlers of the Roe Deer are sometimes crippled by hormonal disorders. They are called 'wigged' Roe Deer

340

341

170 to 275 cm; the tail is 20—38 cm long, shoulder height is 110—155 cm, and the weight is 120—325 kg. The antlers are strong and very knobbly, but regularly of the three-tined type; their size, like that of the animal, depends on the animal's origin: the bulkiest Sambars live in central and northern India (333) and Sri Lanka. The record length of antlers is 129.2 cm from a specimen shot in central India. The Sambar prefers forests and dense bamboo jungles; it is mostly solitary. Gestation lasts 240—260 days; the litter comprises usually one, rarely two young.

Barasinga Deer (334)
Rucervus duvauceli
Cervidae

lives in central and north-eastern India and in Assam, in alluvial regions, dense jungles and at forest verges. According to the season, it lives either solitarily or in quite large herds. Adults reach 170—185 cm in length, the tail is 14—20 cm, shoulder height is 105—115 cm, and the weight is 150—250 kg. Gestation varies from 240 to 250 days; a single young, less often two, are born.

Fallow Deer (335)
Dama dama
Cervidae

is nowadays abundant in European forests, although many people do not realize that it was artificially introduced into a major part of its present territory in ancient times and the Middle Ages. In the post-glacial period,

342

the original home of the Fallow Deer was in the forests of Asia Minor and western Asia. The Fallow Deer prefers park-like open forests, where it lives in herds of up to several dozen head. Its colouration varies with the seasons: the spotted summer coat is replaced in winter by a uniform, grey-brown to light reddish brown coat. The body measures 130 to 145 cm, the tail 14—18 cm, height at the shoulder is 80—85 cm, weight is 50—90 kg. The Fallow Deer's antlers are shaped differently in comparison with other deer species: they are distinctly palmately flattened in the upper parts. Gestation lasts 215—230 days; the female gives birth usually to one young; but even three young may occur very rarely.

Sika (336) lives both in parks and in the wild in many European countries but also in Madagascar,
Cervus nippon
Cervidae

Australia, New Zealand and elsewhere, although it is native to the Amur-Ussuri region, Manchuria, Korea, eastern and south-eastern China and Japan. Its coat is spotted in summer and uniformly brown to brown-grey in winter. The body measures 130—160 cm, the tail 15—20 cm, shoulder height is 85—105 cm, and the weight is up to 125 kg. The Sika has been nearly exterminated in the Amur-Ussuri region and Manchuria, but it is kept as a half-wild animal of economic importance: it provides a substance valuable in the manufacture of medicines — the pantocrine, obtained from dried, immature horns. The female is pregnant for 220 to 230 days and delivers usually one, rarely two, young. In Europe, the Sika readily crossbreeds with the Red Deer in the wild.

Red Deer, Wapiti or **Elk** (337, 338) is the best-known deer, with the most extended area of
Cervus elaphus
Cervidae

distribution of all members of the family, even though it has been driven from many places by the advance of civilization. Its original range included the Atlas Mountains in northern Africa, Corsica, all of Europe, vast part of Asia between 25°—35°N and 50—55°N, and an extensive area in North America from the southern United States to 60°N in Canada. Red Deer naturally form many geographical subspecies in this vast territory: they differ in colouration, size, and shape of antlers. Many of these forms, for example the North American Wapiti Deer or Elk, used to be taken for independent species. Large specimens of Red Deer measure 220—260 cm, the tail is 12—23 cm long, height at the shoulder is 125—150 cm, and the weight is 150—350 kg, reaching 480 kg in North American individuals. Colouration of the coat is influenced by geographical and individual variability, and by the seasons; it can vary from dark grey-brown to almost black-brown, or to pale reddish brown. For decades, rare White Deer (338) have been successfully bred in the Žehušice Deer Park in Bohemia. They represent a white phase of deer imported probably from the Caucasus. The size and massiveness of the antlers vary under similar influences. The heaviest forms of the Eurasian Red Deer have antlers up to 120 cm long, weighing up to 14 kg, while the Wapiti Deer have antlers reaching 160 cm, and weighing more than 25 kg. Red deer are solitary or live in herds, seeking tall thickets and forests. Gestation lasts mostly 240—255 days; a single young is born, less often twins.

Roe Deer (339) is another well-known inhabitant of Eurasian forests. It inhabits almost all Europe
Capreolus capreolus
Cervidae

except northern Scandinavia, and occurs also in the Caucasus, in the forests of southern Siberia, in the Amur-Ussuri region, Manchuria, and in eastern and south-eastern China. Its colouration is grey-brown to reddish brown. The body length varies from 120 to 150 cm, the tail is 2–4 cm long, height at the shoulder is 65–90 cm. The weight averages 15–50 kg. The Roe Deer prefers forests, but it can also be found in more open locations, bush and open woods. After 275–285 days of gestation, the female usually gives birth to two young.

The family of giraffes was represented in the late Tertiary and early Quaternary periods by many species in Eurasia and Africa. Nowadays, they are confined only to Africa and comprise only two species – the Okapi and the Giraffe. They have long to very long necks, and long and slender limbs terminated by pointed, hard hoofs. Their horns, present in males only (Okapi) or in both sexes (Giraffe), are short outgrowths of the frontal or occipital bones; they are never shed. They are covered by hairy skin in the true Giraffe. In adults, the hair becomes worn on the tips of the horns – not the skin, as is sometimes stated. In the Okapi, the tips of horns are bare and often rather sharp.

Okapi (340) is the smaller of the two species of giraffes: it measures 220–255 cm, the tail is 35–45 cm long, the shoulder height is 150–170 cm, and the weight is 220–270 kg. The ground colouration is velvety dark red-brown to reddish black; the limbs are striped white. The Okapi is widespread in eastern parts of the Zaire (Congo) tropical rain forest. It lives solitarily, in pairs, or in small family groups. It feeds mainly on leaves, less often on fruits and seeds of various forest plants. Gestation lasts approximately 435–445 days; a single young is born always. The Okapi was described in 1899 by Governor Johnston of Uganda but the first vague reports of this animal were obtained by the famous explorer Henry Stanley in 1890.

Okapia johnstoni
Giraffidae

Giraffe (341, 342) is indigenous to savanna and forest-savanna regions of almost all Africa south of the Sahara. It has been exterminated in many places. Giraffes have never inhabited sites covered by thick forests, or particularly arid zones.

Giraffa camelopardalis
Giraffidae

343

344

The Giraffe is substantially larger than the Okapi. Body length varies between 370 and 440 cm, the tail is 75 – 110 cm long, height at the shoulder is 250 – 350 cm, and the maximum height to the crown of the head may exceed 580 cm. The weight of large specimens is about 750 to 1,600 kg. Several geographical races (subspecies) are distinguished according to colouration and certain anatomical features; the most outstanding is the Reticulated Giraffe (*Giraffa camelopardalis reticulata*) (341).

345

346

Rothschild's Giraffe (*Giraffa camelopardalis rothschildi*) (342) is another distinct subspecies. Giraffes are found chiefly in herds of ten to 15, but exceptionally large herds of more than 70 have also been observed. The female is pregnant for 430—455 days; she gives birth to one, exceptionally two young.

347

348

Pronghorn Antelope (343) is the only member of the pronghorn family, native to the vast grasslands of North America, from south-western Canada, over the western United States to northern Mexico. It has become extinct in many places: its numbers were drastically diminished by white hunters throughout the last century. In external appearance, the Pronghorn resembles the antelopes. Colouration of its coat is reddish-brown above and white below and around the tail. The head and neck have pronounced light and dark markings. The horns are the most remarkable feature of pronghorns: in structure, they are true horns, occurring in both sexes but, unlike the true hollow horns of bovids (Bovidae), they are forked and shed regularly. Evolutionarily, the Pronghorn appears to be an advanced, entirely independent type of ruminant. Its body measures 110–150 cm, the tail is 10–17 cm long, height at the shoulder is 85 to 105 cm, and the weight is 36–65 kg. Pronghorns live in small groups of ten to 15 specimens in summer, and in winter gather in herds numbering one to several hundred head. Old males are sometimes solitary. The Pronghorn is the fastest New World mammal: over a short distance, it can reach a speed of more than 75 km/hour. The female undergoes 230–240 days of gestation and usually gives birth to two young.

Antilocapra americana
Antilocapridae

The last and most numerous family of artiodactyls is that of bovids. Bovids are a diverse group, including both small and large artiodactyls. They are found in North America, Africa, Eurasia, Indonesia, and in the Philippines. Their limbs have more or less stunted second and fifth digits. Horns in bovids are formed by bony outgrowths of the frontal bone, covered by horn. The horns are permanent growths, lasting virtually a lifetime, and usually present only in males; if they occur in females, they are less developed. All species of the bovid family have one pair of horns, except the Indian Four-horned Antelope (*Tetracerus quadricornis*), which has two pairs of short horns. Bovids usually produce a single young.

349

Black-fronted Duiker (344) is the first, although not very typical member of the family. It lives in thick virgin forests of central Africa, is about 50 cm tall, and weighs 14—18 kg. Gestation lasts 118—125 days; a single young is generally born.

Cephalophus nigrifrons
Bovidae

Yellow-backed Duiker (345) is the largest duiker. Its body reaches 115—154 cm in length, its height at the shoulder is 68—85 cm, and its weight is 40—50 kg. The coat is velvety dark brown, with an orange or rust-coloured head tuft and a wedge-shaped, light orange stripe on the back. The Yellow-backed

Cephalophus sylvicultor
Bovidae

350

351

Duiker lives in primary forests of western and central Africa. Gestation lasts 120—135 days; generally a single young is born.

Kirk's Dikdik (346) is one of the smallest antelopes; the length of the body is 55—65 cm, shoulder height is 37—45 cm, and the weight is about 3—4.5 kg. The small horns, often concealed in a tuft of longer hair on the crown of the head, are present only in males. The Dikdik's main feature is the nose, elongated into a short proboscis. Kirk's Dikdik frequents the bush and steppe thickets of south-west and eastern Africa. In relation to its smallish size, the duration of pregnancy is uncommonly long — about 175—185 days. The litter regularly consists of a single young.

Madoqua kirki
Bovidae

Klipspringer (347) is a smallish, pepper-and-salt-coloured antelope, widespread in the rocky hills and mountains of Africa, from the Sudan, Ethiopia and Somalia to the southernmost parts of the continent, and to south-western Africa. It also lives in a small area of northern Nigeria. The Klipspringer has strong, tall hoofs, touching the ground only with their tips. This enables

Oreotragus oreotragus
Bovidae

353

the animal to move with agility and leap on rocky ridges literally no wider than a few centimetres. The straight, short horns usually occur only in males. Gestation lasts 210—220 days; as in the majority of antelopes, the litter contains usually a single young.

Bushbuck or **Harnessed Antelope** (348) inhabits the brushwood and bush of most of Africa south of the Sahara. Its body measures up to 155 cm, the height at the shoulder is 70—95 cm, and it weighs up to 80 kg. Horns may reach 57 cm in length; they are absent in females. The period of gestation averages 225—230 days.

Tragelaphus scriptus
Bovidae

Sitatunga (349) is related to the preceding species. It inhabits swamps, bogs and humid rain forests of western and central Africa. It is the only really 'amphibious' antelope. It has banana-shaped, elongated hoofs, which prevent it from sinking into mud. The Sitatunga is larger than the Bushbuck: it stands 100 to 110 cm high at the shoulder, and weighs 125—130 kg. The horns are usually 60—65 cm long, and are present only in males. Following 245—258 days of gestation, a single young is born.

Tragelaphus spekei
Bovidae

Lowland Nyala (350) is approximately the same size as the Sitatunga. It inhabits south-eastern Africa from Natal to Malawi. Living in small groups, it seeks chiefly lowlands covered by a thick bush, and wooded grasslands. It is not particularly restricted to humid sites, but requires to be near water. Horns, up to 80 cm long, are present only in males, which are also distinguished by a mane of long hair on the neck, throat, back and abdomen. Males are dark grey-brown with purplish overtones, females are usually chestnut brown. Gestation lasts 220—240 days.

Tragelaphus angasi
Bovidae

354

Lesser Kudu (351) lives in the bush, in acacia coppices, or in thick, semi-arid brushwood locations of Somalia, Ethiopia, the southern Sudan, and eastern Africa. Body length averages 150—175 cm, the tail is about 25—30 cm long, shoulder height is 95—105 cm, and the weight is 90—115 kg. Horns, present only in males, have a wide spiral twist, attaining a length of up to 90 cm. The Lesser Kudu usually stays in pairs or families; females sometimes gather in small, independent herds. It is a shy, relatively rare antelope species.

Tragelaphus imberbis
Bovidae

Greater Kudu (352, 353) is a bigger relative of the Lesser Kudu and is the third largest existing antelope. Its body reaches 220—270 cm in length, a shoulder height of 135—160 cm, and a weight of 270—330 kg. Females are smaller, weighing 280—210 kg. Horns are present only in males: they are massive, up to 171.5 cm long. The Greater Kudu inhabits most of southern Africa south of the Zambezi; it penetrates westwards to Angola and ranges

Tragelaphus strepsiceros
Bovidae

355

356

across all eastern Africa north to the Sudan and Ethiopia. Like the Lesser Kudu, it avoids open grasslands, preferring thickets, bush and open forests. Besides size, the two species differ in other features: the Greater Kudu has a conspicuous mane on the throat, neck and back, while the Lesser Kudu has slightly longer hair only on the neck and front part of the back. Two pronounced white spots on the Lesser Kudu's throat are

357

358

lacking in the Greater Kudu, which usually also has eight to ten narrow, crosswise white stripes on the flanks; the Lesser Kudu has at least eleven, but usually twelve to 15 of these. The Greater Kudu lives mostly in small groups of four or five individuals; old males are sometimes solitary, or stay in independent male groups. Gestation lasts about 210—245 days in both species; almost without exception, a single young is born.

359

360

Cape Eland (354) is, with the Giant Eland, the largest living antelope species. Body length in old males reaches 260—320 cm, the tail is 60—90 cm long, height at the shoulder is 150—175 cm, and the weight is 600—900 kg. Horns, present in both sexes, are bulkier in males, but not always longer than in females. Strong specimens have horns 60—70 cm long; the maximum length recorded is 110.5 cm. The Cape Eland lives in small or large herds, in grasslands and forested savanna to altitudes over 3,000 metres in a large part of Africa from the southern Sudan to Cape Province, south-western Africa and southern Angola. Old males have a colossal and imposing appearance, reminding one more of the bulls of wild oxen than of antelopes. Females are pregnant for 255—270 days; a single young is the rule.

Taurotragus oryx
Bovidae

Giant Eland (355) may be even bigger than the Cape Eland. Old males can measure up to 425 cm, including the tail, stand 182 cm at the shoulder, and can weigh exceptionally 950—1,000 kg. Horns in males are always heavier and longer than in females. Large specimens have horns 70—80 cm long; the maximum reported length is 121.2 cm. The Giant Eland differs from the Cape Eland in having a bolder and more dense pattern of crosswise white

Taurotragus derbianus
Bovidae

361

stripes on the flanks (this is sometimes completely absent in the Cape Eland), a longer mane on the throat and neck, a longer dewlap on the throat, and more massive and wider spaced horns. It prefers forests and dense bush. It feeds more on leaves of bushes and young trees than on the grass which the Cape Eland prefers. It lives mostly in small groups and is widespread in western Africa, mainly in the area of the upper reaches of the Niger and Bani Rivers, in central Africa from Nigeria to the southern Sudan, Uganda, and western Kenya. Gestation corresponds to that of the Cape Eland.

Bongo (356) lives solitarily, in pairs or in small groups in the virgin forests of western and central
Boocercus euryceros
Bovidae

Africa, from Sierra Leone to western Kenya. Its body reaches 240 cm in length, the tail is 50 – 65 cm long, the height at the shoulder is 115 to

362

The Bluebuck, like the Cape Lion and the Quagga, was a large south-African mammal totally exterminated by Man

363

130 cm, and the weight of old individuals is 170—235 kg. Horns are present in both sexes, but are smaller in females. Gestation is estimated at 8½ months; a single young, occasionally twins, are dropped.

Roan Antelope (357)
Hippotragus equinus
Bovidae

has a body up to 255 cm long, a tail 50 cm long, a shoulder height up to 160 cm and it weighs 220—280 kg. Males and females are the same colour, and both have horns, which are thinner and shorter in females. Horns in the males average 60—75 cm in length; the maximum length recorded is 99.1 cm. The Roan Antelope prefers open locations, but avoids completely open savanna. It is widespread from western Africa, the Sudan and Ethiopia, across eastern Africa to Bechuanaland and Transvaal. Gestation lasts 270—285 days; usually one young is born.

364

365

Sable Antelope (358) inhabits open forests, forested savanna and the bush from southern Kenya to the Transvaal and eastern Angola. The body measures up to 245 cm, the tail 40—50 cm, shoulder height is 130—150 cm, and the weight is 200—240 kg. Horns are present in both sexes: their average length in males is 110—125 cm; the maximum length recorded is 164.6 cm. Males are glossy black, females are brown-coloured. Sable Antelopes live mostly in herds of ten to 20 individuals. Their habits and time of gestation are the same as those of the Roan Antelope.

Hippotragus niger
Bovidae

Bluebuck was probably the first African mammal exterminated by white men during their expansion and colonization of southern Africa. The Bluebuck used to be indigenous to the south-western part of Cape Province. The last specimen of the species was killed in 1799 or 1800. Its ground colouration was bluish grey, off-white below, brownish on the forehead, and light cream to whitish around the eyes and on the face. The Bluebuck was smaller than the Sable Antelope; horns in both sexes usually did not surpass 45—50 cm in length.

Hippotragus leucophaeus
Bovidae

Nilgai or **Blue Bull** (359) is the largest Asian antelope. The body is usually 190—230 cm long, the tail measures 45—55 cm, height at the shoulder is 120—130 cm, and the maximum weight is 190—210 kg. Females are smaller by approximately one-fifth and have no horns. Horns in the males reach 15—20 cm (25 cm is the maximum length recorded). The Nilgai's haunts are the forests, jungles and savanna of the Indian peninsula. It lives in small herds; old males are sometimes solitary. After 240—250 days of gestation, the female drops one or two young.

Boselaphus tragocamelus
Bovidae

366

Addax (360) is nowadays one of the rarest antelopes. It used to inhabit the whole Sahara region from Mauritania, Morocco, and Algeria to the Sudan. Scattered islands of desert vegetation provide its food; its requirement of water is more or less satisfied by water contained in the plants. The Addax is a comparatively heavily-built antelope, with wide hoofs which prevent it from sinking into sand. It reaches 150—175 cm in length, the tail is 27—35 cm long, height at the shoulder is 95—112 cm, and its weight is 65—135 kg. Long, relatively heavy horns are present in both sexes; in males, their length reaches 70—85 cm; the maximum length recorded is 109.2 cm. The Addax has a characteristic dark brown-and-white facial mask; some specimens are almost white, mainly in summer. The Addax lives mostly in small herds of 15 to 20 head: their migration in the desert is determined by the rare and irregular rainfall which literally draws out from the sand the vegetation on which the animals feed. Gestation is very long: following 310—340 days, a single young is born.

Addax nasomaculatus
Bovidae

Scimitar-horned Oryx (361) is a stout antelope, occurring in the same localities as the Addax. While the Addax penetrates into true deserts, the Scimitar-horned Oryx is found only on desert margins, in semi-deserts, and dry grasslands. It does not occur in grassy savannas. Its body is 190—220 cm long, the tail 45—60 cm, shoulder height is 110—120 cm and the weight is 150 to 210 kg. The curved horns are always more slender, but sometimes longer in females than in males. They can reach 80—100 cm; the maximum length known is 127.3 cm.

Oryx dammah
Bovidae

Gemsbok or **Beisa Oryx** (362) is of the same size as the Scimitar-horned Oryx, but it has almost straight horns, present in both sexes. It is resident in dry savanna and the bush of north-eastern, eastern and south-western Africa. It lives

Oryx gazella
Bovidae

367

mostly in groups of 5 to 40 head. In both oryx species pregnancy lasts approximately 265—275 days; usually a single young is born.

Bontebok or **Blesbok**
Damaliscus dorcas
Bovidae

(363) is a rare bovid occurring in isolated places in the South African grasslands. Its body measures 160—185 cm, the tail is 25—40 cm long, height at the shoulder is 85—100 cm, and the weight is 80—110 kg. The

368

369

horns are 30—35 cm long; the maximum length recorded is 50.5 cm. Those of females are shorter and more slender. Gestation lasts 220—240 days; almost invariably a single young is born.

Hunter's Hartebeest (364) is the rarest antelope of the hartebeest group. It is confined to a small region of the Kenya-Somalian border, inhabiting arid savannas and bush. Its body measures 170—205 cm, its tail is 30—35 cm long, its height at the shoulder is 95—110 cm, and its weight is 60—85 kg. The forehead is marked from eye to eye by a characteristic narrow stripe; the tail is white-tipped. Horns are up to 60 cm long; the maximum length is 72.4 cm. They are smaller in females. Hunter's Hartebeests live in herds of ten to 25 head. Gestation lasts 210—240 days; usually a single young is dropped.

Damaliscus hunteri
Bovidae

Sassaby or **Swift Topi** (365) is relatively abundant in open savanna and park-like forests from western Africa to the Sudan, and across eastern Africa to northern Transvaal and northern Natal. It is larger than the two preceding species. Horns are present in both sexes: their length is 40—50 cm; the maximum length recorded (in the subspecies *Damaliscus lunatus korrigum*) is 72.4 cm. The Sassaby lives mostly in groups of ten to 30 individuals, but these may form herds of several hundred or thousand animals. The female is pregnant for 230—245 days; usually a single young is born.

Damaliscus lunatus
Bovidae

Red Hartebeest (366, 367) is quite a common antelope species. It is found in Africa in several geographical races, like the Sassaby. It differs from the three species described above in various features; it has a different shape to the head and shorter, typically twisted horns, present in both males and females.

Alcelaphus buselaphus
Bovidae

The Red Hartebeest is the largest hartebeest species, standing up to 125–130 cm at the shoulder, and weighing even more than 200 kg. It is widespread in savanna, in the bush, and in semi-arid grasslands from western Africa to the Sudan and Somalia, in eastern, southern and south-western Africa. The various geographical forms include Coke's Hartebeest (*Alcelaphus buselaphus cokii*) (366) from Kenya and Tanzania, and Lichtenstein's Hartebeest or Konzi (*Alcelaphus buselaphus lichtensteini*) (367). The Red Hartebeest lives in herds of 10 to 30, but sometimes it forms herds of several hundred. Gestation lasts 214–242 days; most often a single young is born.

Brindled Gnu or **Wildebeest** (368, 369) occurs in several geographical races in savanna, in the bush and in some of the park-like forests of eastern, southern, and south-western Africa. Its greyish-coloured body is 185–215 cm long, the tail measures 40–55 cm, the height at the shoulder is 115–122 cm, and the weight is 150–250 kg. The pointed, outward-curved horns are present in both sexes. The Blue Wildebeest (*Connochaetes taurinus taurinus*) (368) is the most common subspecies, living in southern parts of the range of distribution. The White-bearded Gnu (*Connochaetes taurinus albojubatus*) (369) inhabits the savanna and steppes of Tanzania. Brindled Gnus live mostly in large herds of several dozens to thousands of head. Females undergo 245–255 days of gestation and, as a rule, drop a single young.

Connochaetes taurinus
Bovidae

White-tailed Gnu (370) was native to grassland in the heart of southern Africa. Nowadays, it is restricted to a few wildlife reserves and to the pastures of several large private farms. The White-tailed Gnu is somewhat smaller than the

Connochaetes gnou
Bovidae

370

preceding species; it weighs 130–180 kg. The ground colour of its body is deep brown to black-brown, and the tail is white. The appearance of this species is even more bizarre than that of the Brindled Gnu, owing to an erect tuft of hair on the muzzle, and to the forward- and upward-curving, pointed horns, present in both sexes. Horns in both species are

371

372

roughly of the same length (50—65 cm); the maximum length of horns recorded is 83.8 cm in the Brindled Gnu and 79.6 cm in the White-tailed Gnu. Herds of White-tailed Gnu number mostly 20 to 30 head. Gestation lasts 250—265 days, and a single young is almost always born.

373

Bohor Reedbuck (371) inhabits savanna and open bush from western Africa to the Sudan and Ethiopia, as far south as southern Tanzania. It is a medium-sized antelope with rather short horns, present only in males. Reedbucks live in pairs or in families, sometimes also solitarily. Gestation averages 225–235 days; one or two young are born.

Redunca redunca
Bovidae

Waterbuck (372) lives predominantly in open forests or bush and in alluvial regions from Senegal and Guinea to the Sudan and western Ethiopia, in eastern Africa, and southwards approximately to the Tropic of Capricorn. Its body measures 180–225 cm, the tail is 22–40 cm, its height at the shoulder is 115 to 137 cm and its weight is 180–250 kg. Horns are present only in males, and average 60–70 cm in length; the maximum length recorded is 99.8 cm. Two forms are distinguished: the Kringgat Waterbuck (*Kobus ellipsiprymnus ellipsiprymnus*) with a pronounced, elliptic white ring around the tail, and the Defassa Waterbuck (*Kobus ellipsiprymnus defassa*) (372), without the ring on the hindquarters. The former subspecies lives in the south-eastern parts of the range of distribution, the latter form inhabits the western and north-western section, and central Africa. Pregnancy lasts 210–240 days; one, exceptionally two young are born.

Kobus ellipsiprymnus
Bovidae

374

Kob (373) is somewhat smaller than the preceding species: body length usually does not exceed 185 cm; the weight is 85–118 kg. Horns, usually 45–50 cm long, are present only in males; the maximum length recorded is 65 cm. The Kob frequents bushy steppes and savanna, never too far from water, and alluvial river basins from Senegal and Gambia to Chad, the south-western Sudan, northern Zaire, and Uganda. It exists in small herds of five to 30 individuals. Gestation lasts 210–230 days; one or two young are born.

Kobus kob
Bovidae

Lechwe (374) is very similar to the Kob both in size and appearance. It is found in forests, alluvial river regions, and swamps, mainly in southern parts of Zaire, Zambia, and eastern Angola. It forms medium or large herds of several hundred head. The Lechwe seeks food both on land and among aquatic vegetation, wading in water as deep as 70 cm. Pregnancy lasts 225–245 days; usually a single young is born.

Kobus leche
Bovidae

Impala (375) is resident in forest grasslands, bushy forests, and on the edge of savanna from Kenya to southern Africa. It is a lightly-built, medium-sized antelope, standing 77–95 cm at the shoulder, and weighing 50–75 kg. Horns, 50–60 cm long, are present only in males; the maximum length is 91.8 cm. The Impala lives in large herds or smaller groups of 15 to 25

Aepyceros melampus
Bovidae

375

head, composed of females led by a sole male. Impalas are noted for their graceful, extremely long and high leaps. Females undergo 170 – 190 days of gestation, giving birth to one, less often to two young.

Gerenuk or **Giraffe Gazelle** (376, 377) lives in Tanzania, Kenya, Ethiopia and Somalia; it prefers bush, arid places covered with thorny shrubs, or open, park-like acacia woods. It is a long-legged, slender antelope with an extremely long neck. It eats leaves of trees and shrubs: trying to reach higher branches and twigs, the Gerenuk stands up in a typical pose on its hind limbs (376). It lives solitarily or in small groups of three to seven individuals. Horns are relatively bulky and 30 – 35 cm long; the maximum length recorded is 43.6 cm. They are present only in males. Gestation takes 7 – 8 months; generally a single young is dropped.

Litocranius walleri
Bovidae

Thomson's Gazelle (378) is restricted to grassy savannas of eastern Africa, where it is one of the most abundant species. It stands 60 – 65 cm at the shoulder and weighs 17 – 26 kg. The slender, slightly curved horns occur in both sexes, being smaller in females. The average length of the horns is 30 to 35 cm; 43.4 cm is the maximum known. Thomson's Gazelles live most frequently in harem herds of several dozen head with only one adult male. The herds are led, however, by a senior female. Gestation lasts 155 – 175 days; one or two young are usually born.

Gazella thomsoni
Bovidae

Persian Gazelle or **Dzheyran** (379) is the Asiatic representative of the gazelles. It has become extinct or extremely rare in many places. Originally, it ranged from Iran, Afghanistan and Pakistan, across central Asia, to Mongolia and central China. Its body is 100 – 120 cm long, the tail measures 16 – 20 cm, and its shoulder height is 65 – 75 cm. The horns are 25 to 35 cm long; the maximum length recorded is 41 cm. Persian Gazelles live in steppes and semi-deserts, but also in arid mountain valleys, often at altitudes of more than 3,500 metres. In summer, they form groups of two to five, and in winter they gather in herds of several dozen to hundreds of head. Gestation lasts 165 – 180 days; usually two young are born.

Gazella subgutturosa
Bovidae

Grant's Gazelle (380) lives predominantly in the territory of Kenya and Tanzania, settling in savanna, wooded savanna and the bush. It usually stays in groups of six to 30 individuals. It is larger than Thomson's Gazelle: the height at the shoulder is 75—85 cm, the weight 45—75 kg. The slender, slightly lyre-shaped horns measure 55—65 cm in males, 80.5 cm at the most. In females they are shorter and more slender. Gestation lasts 165—178 days; most often a single young is born.

Gazella granti
Bovidae

Springbok (381, 382) is a typical inhabitant of the open, grassy steppes, the bush, and the plateau of southern and south-western Africa. A medium-sized animal, it reaches 120—145 cm in length, 73—84 cm at the shoulder, and it weighs 32 to 45 kg. The lyre-shaped horns are more slender and straighter in females; in males they measure 30—35 cm, 48.5 cm at maximum. Springboks are famous for their ability to bounce into the air. The characteristic leaps they make when surprised, endangered or simply playing, may reach up to 3—3.5 metres in height and are repeated five or six times in succession: during this stiff-looking action, the back is arched, head lowered, and the legs held fully extended with the hooves almost bunched together. A crest of long, white hair appears simultaneously on the back; when the animal is calm, the crest is down, partly concealed by the dark hair. Gestation

Antidorcas marsupialis
Bovidae

376

377

378

lasts 170—185 days; the female gives birth to a single young, less often to twins.

Saiga (383) has a characteristic stubby nose, elongated into a sort of proboscis, strangely magnifying the already large and heavy head. The Saiga originally ranged from the Russian steppes, across central Asia to Sinkiang and Mongolia.

Saiga tatarica
Bovidae

379

380

It has become extinct in the western part of the territory, but not totally so: in some places, its numbers are quite substantial. It is medium-sized, weighing 25—51 kg. Horns are yellowish to yellow-grey, 25—30 cm long, with a known maximum of 38 cm, and present only in males. Saigas mostly stay in herds of ten to 40 head. Gestation lasts 140—147 days. The females assemble at the time of delivery into larger groups, always in specific, fixed locations, where they usually give birth to twins.

Blackbuck (384) is another Asian species of antelope. It ranges from Pakistan to Bengal, and from the slopes of the Himalayas to the southernmost part of the Indian Peninsula. It avoids hilly and forest regions, preferring open country. Its body measures 100—130 cm, the tail is 14—17 cm long, the height at the shoulder is 65—75 cm and the weight is 30—40 kg. The spirally-twisted horns, reaching 68.5 cm at maximum, occur only in males.

Antilope cervicapra
Bovidae

382

381

Colouration differs in the two sexes: males are deep brown to black-brown, females are fawn. Underparts are always white or off-white. Blackbucks live mostly in herds of ten to 50 head. The female is pregnant for 170 to 180 days, and bears one or two young.

Goral (385), like the Chamois, is allied more to goats and sheep than to antelopes, as used to be claimed. It is widespread from the western Himalayan slopes, along the whole Himalayan range to south-eastern China, and northwards to the Amur-Ussuri region. It prefers mountainous and rocky locations neighbouring upon forests, and situated at altitudes of 1,000—2,700 metres. Its body measures 95—125 cm, the tail is 10—18 cm long, the height at the shoulder is 55—70 cm, and the weight is 20—35 kg. The horns curve slightly backward, and occur in both sexes; they are 12—18 cm long. Gorals usually live in families of two to eight members. The female is pregnant for 170—185 days, and bears a single young, rarely two.

Nemorhaedus goral
Bovidae

Chamois (386) inhabits rocky regions of Europe and Asia Minor. It is somewhat larger than the Goral; the slender, backward-hooked horns are present in both sexes. They are 15—22 cm, exceptionally 34 cm long. The Chamois can be seen mainly on scarcely accessible rocky mountain slopes, at altitudes up to 4,000 metres. It lives in groups of 15 to 30 members; old males are often solitary. Gestation lasts usually 160—170 days; a single young is usually born, but sometimes twins and exceptionally triplets.

Rupicapra rupicapra
Bovidae

Rocky Mountain Goat (387) is a long-haired, white-coloured, and comparatively large North American relative of the Asian Goral and Eurasian Chamois. It inhabits the whole region of the Rocky Mountains and some other mountain ranges in the western United States, from Alaska and the Yukon to Montana, central Idaho and northern Oregon. It was also introduced to the Black Hills of South Dakota. Horns in the males are 20—25 cm,

Oreamnos americanus
Bovidae

383

exceptionally 32 cm long; females have narrower and shorter horns. Mountain goats climb nimbly on snow-covered rocky cliffs, sometimes occurring at altitudes above 4,000 metres. The female is pregnant for 168—180 days, and gives birth to one or two young.

Wild Goat (388), with other species of wild goat, ibex and sheep, inhabits mountainous and rocky locations. All these animals move in their environment with the absolute skill of perfect climbers. The Wild Goat used to live in south-western, mountainous areas of Asia, from Turkey over the Caucasus and Iran, to central Asia, Afghanistan and Pakistan. It was also found on Crete, and on the small island of Erimomilos in the Mediterranean Cyclades. It has become rare in most of its original territory, but it has been introduced into various European localities. Body length of old males, which are larger than females, is 130—150 cm, the tail is 15—20 cm long, height at the shoulder is 75—90 cm, and the weight is 35—65 kg. The arching horns of males are 80—130 cm long, while females have horns only 20—30 cm long, much narrower and straighter. Wild goat (and ibex) live mostly in groups of ten to 20 animals. Old males are often solitary. Gestation lasts 155—170 days on average; females bear one or two young.

Capra aegagrus
Bovidae

Mountain Ibex (389) inhabits the Alpine mountain range, occurring at altitudes over 3,500 metres. From here, it was introduced to mountains in other parts of Europe (for example the Carpathian Mountains). Its original home also includes

Capra ibex
Bovidae

384

rocks and mountains of north-eastern Sudan, eastern Egypt, and the Arabian peninsula. Two geographical subspecies are distinguished: the Alpine Ibex (*Capra ibex ibex*) (389), and the Afro-Arabic or Nubian Ibex (*Capra ibex nubiana*). The Mountain Ibex stands as much as 95 cm high at the shoulder; the body is 160—165 cm long. The weight of old males is 70—80 kg; females are substantially smaller. Horns are present in both sexes but, as in other wild goats, ibexes and sheep, they are shorter and more slender in females. The length of horns in old bucks can exceed 100 cm.

Siberian Ibex (390) is
Capra sibirica
Bovidae

sometimes regarded as a mere subspecies of the Mountain Ibex. It is resident in the mountain ranges of central Asia, in the Pamirs, Altai, westernmost Himalayas and Mongolian mountains, at altitudes above 4,500 metres. Unlike the Mountain Ibex, it sometimes descends below

385

the timberline, to altitudes of 450—600 metres. The Siberian Ibex stands up to 100 cm at the shoulder, and weighs 60—100 kg. Horns in the males often exceed 100 cm; the maximum length recorded is 147 cm.

Caucasus Ibex or **Caucasus Tur** (391) is confined to a relatively small territory of western
Capra caucasica
Bovidae

Caucasus at altitudes of 800—3,500 metres. It is of the size of the Siberian Ibex, but its horns are much shorter (65—70 cm), although they are relatively heavier.

Dagestan Ibex or **Dagestan Tur** (392) inhabits the eastern Caucasus; only a smaller part of
Capra cylindricornis
Bovidae

central Caucasus is frequented by both the Caucasus Ibex and this species. The Dagestan Ibex is found chiefly at altitudes of 2,000—4,200 metres. It reaches approximately the size of Siberian and Caucasus Ibexes, although old bucks are heavier, and weigh more than 110 kg. Horns in old males are extremely massive, but lack the prominent protuberances on the surface, which are typical of the horns of the preceding species of ibex and the Wild Goat. The horns have a unique shape, resembling a backward-tilted lyre. Their length often exceeds 80—90 cm; 103 cm is the maximum length. The base perimeter of the largest horns in the Mountain and Siberian Ibexes is 25—28 cm, in the Caucasus Ibex 28—30 cm, while in the Dagestan Ibex it is 35 cm or more.

Markhor (393) lives in the Pamirs, Hindukush and western Himalayas, in northern Afghanistan,
Capra falconeri
Bovidae

Pakistan, and in the northermost parts of India, at altitudes of 1,500 to 3,500 metres. The Markhor is as large as the Siberian Ibex, but it is more lightly built: it weighs 70—85 kg. Horns in old males can often surpass 100 or even 130 cm, if the length of the spiral is measured; 159.5 cm is the maximum length recorded. The Markhor occurs chiefly in small groups or individually.

386

387

Barbary Sheep or **Aoudad** (394) frequents mountains and rocks from the northern bend of the Niger to the Sudan and Ethiopia. Beside the Nubian form of the Mountain Ibex and the Ethiopian Ibex (*Capra wallie*), it is the only representative of wild goats and sheep on the African continent. The Barbary Sheep is a stout animal, fawn-coloured, and living in groups of five to 20 individuals. Males, heavier than females, reach 190 cm in length, including the 20—22 cm long tail. Height at the shoulder is 100 cm, and the weight may reach 115 kg. Horns in males are 50—65 cm long; 86.9 cm at the most. Gestation lasts 154—161 days; the litter numbers one or two young.

Ammotragus lervia
Bovidae

Mouflon (395, 396) is the only indigenous wild sheep of Europe. It is likely that the Mouflon was originally widespread in many south-European mountain ranges. It is a long time, however, since it was forced to withdraw to its last refuge:

Ovis musimon
Bovidae

388

389

390

391

the Mediterranean islands of Sardinia and Corsica. Since the 18th century, it has been introduced to many European countries, where it is relatively abundant nowadays. The greatest numbers of Mouflons are acknowledged to occur in Bohemia, Moravia and Slovakia. Males are 130–140 cm long, including a tail 10 cm long; height at the shoulder is 70–75 cm; old rams weigh more than 55 kg. Females are smaller, and often lack horns. The huge, coiled horns of males measure 60–80 cm, exception-

392

393

ally more than 100 cm. Mouflons prefer rocky areas and open mountain slopes, but they also thrive in wooded, hilly regions. They live in groups of several dozen. After 148–155 days of gestation, the female drops a single young, less often two.

394

Bighorn Sheep (397) is a robust sheep, ranging from mountain ridges of the Baïkal region across the Trans-Baïkal area, Kamchatka, eastern Siberia, to the Bering Strait, and then from the Alaskan mountains over western parts of Canada and the United States to southern slopes of the Wasatch Range in Utah. Old rams can measure up to 195 cm, stand 100—103 cm high at the shoulder, and weigh 150—175 kg. Their horns can be over 110 cm long and remarkably strong. Females are smaller and have only short, slightly curved horns. The Bighorn Sheep exists in many subspecies, differing only in size, robustness, colour, and shape of horns. Colouration varies from dark brown-grey (for example in the Kamchatka Sheep, *Ovis canadensis nivicola*) to cream white in the Alaskan Sheep (*Ovis canadensis dalli*) (397). The Bighorn Sheep inhabits the most impenetrable mountain slopes and valleys, living in groups averaging 15—20 individuals. Gestation lasts 170—180 days; the litter comprises one or two young.

Ovis canadensis
Bovidae

Argali or **Central Asiatic Wild Sheep** (398) is the largest wild sheep. It is resident in the Asian mountain ranges from Asia Minor across Iran, central Asia, Afghanistan, the Himalayas, to the Trans-Baikal region and northern Manchuria. Throughout this territory, it occurs in many geographical forms. (Some zoologists are of the opinion that the European Mouflon is just another subspecies of the Argali.) One of its very large geographical forms is the Marco Polo Argali (*Ovis ammon polii*), living in the Pamirs; its horns, up to 190 cm long and twisted in wide spirals, extend far to the side. Males of the largest form, the Altai Argali (*Ovis ammon ammon*) (398), measure up to 215 cm, including the tail 12—17 cm

Ovis ammon
Bovidae

395

396

long; the height at the shoulder is 122–125 cm and the weight is 200 to 230 kg. The Argali prefers mountain plateaux, treeless ridges, and wide valleys of the steppe type, avoiding continuous forests and steep rocky slopes. Argalis live in small groups and herds of up to several dozen.

397

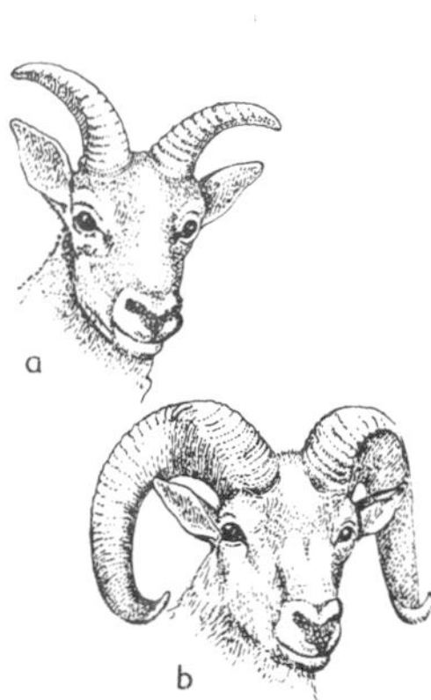

Difference in size and shape of the horns in females (a) and males (b) of the Bighorn Sheep

398

The female is pregnant for 150—160 days, and gives birth mostly to a single young.

Tahr (399) occurs in three mutually isolated and widely-separated places: in the Himalayas, in southernmost India, and in the south-western tip of the Arabian peninsula. Each area is inhabited by a different subspecies. The Tahr frequents mainly mountainous and rocky slopes in the forest zone; the Himalayan Tahr (*Hemitragus jemlahicus jemlahicus*) (399) thus occurs only up to 3,000—3,500 metres above sea level. The Tahr's body measures 120—160 cm, the tail is about 15 cm long, the height at the shoulder is

Hemitragus jemlahicus
Bovidae

399

65—95 cm, and the weight is 40—100 kg. Horns are present in both sexes: in males, they measure 30 cm on average, 45 cm at the maximum. The long hair, forming a paler mane on the neck, is grey-fawn, brown, or slate-grey. The female undergoes 180—240 days of gestation, giving birth to one or two young.

Takin (400) is a little-known representative of Asian ruminants. It is found in the mountainous
Budorcas taxicolor
Bovidae

areas from Bhutan and northern Burma to the mountains of central China in the provinces of Szechuan and Shensi, and in the easternmost parts of the Kansu province. The Takin is a stout, heavily built animal, seemingly combining features of oxen, musk oxen, goats, sheep, and antelopes. Its body measures 170—225 cm, the tail is 15—20 cm long, the height at the shoulder is 100—130 cm and the weight is 200—350 kg. The massive backward-turned horns are present in both sexes; their length can reach 63.5 cm in males. The ground colouration of the long hair varies from yellowish white-grey to brown. The Takin lives in herds in thickets on the timberline at altitudes of 2,000—4,200 metres. The female is pregnant for 200—220 days and usually bears a single young.

400

401

Musk Ox (401) is the only representative of an independent genus of ruminants, inhabiting arctic and subarctic regions of Alaska, Canada, and Greenland. It is relatively rare in some of its original habitats, but it was artificially introduced to central Scandinavia where it does very well. The Musk Ox has a robust body measuring 190–245 cm; the tail is 7–11 cm long, shoulder height 110–145 cm and the weight of large, old bulls reaches 340–410 kg. Its horns (maximum length 73.7 cm) are characteristically shaped, forming a sort of helmet. The Musk Ox has a hump of fat along the neck, and broad, solid hoofs, preventing it from sinking into snow. Musk Oxen live in herds of ten to 35 head, sometimes of several hundred. In case of danger, the herd assembles in a defensive circle, protecting the offspring in the middle. This is also a highly effective stratagem in the deterrence of wolves. Gestation lasts 265–275 days; the litter consists of a single calf.

Ovibos moschatus
Bovidae

Kerabau or **Water Buffalo** (402) is undoubtedly a descendant of the Arni. According to the concepts of classification of species of domestic animals outlined in the Introduction, it is considered to be an independent species, not merely a form of its wild ancestor, as many authors suggest. The Kerabau is somewhat smaller than the Arni, and for many centuries it has been bred not only in southern Asia, but also in Egypt, the Balkans, in the Asian Islands, including the Philippines, and elsewhere.

Bubalus bubalis
Bovidae

Arni or **Wild Water Buffalo** (403) is a typical representative of wild oxen. It is native to southern and south-eastern Asia, from India and Nepal southwards and eastwards to Sri Lanka and the Sunda Islands. It has been totally exterminated in many places. The body length of large bulls is 260–285 cm, the tail is 60–70 cm long, height at the shoulder is 155–170 cm, and the weight is 600–900 kg. The backward-arching, long horns are present in both sexes, as in all oxen, but those of males are more massive and longer, often reaching 140–150 cm; the maximum length recorded is 194 cm. The Arni lives in herds, mostly in regions covered with tall grass, shrubs

Bubalus arnee
Bovidae

402

and sparse woods, always near water. Females are pregnant for about 295—300 days; the litter comprises one or two young.

Gaur or **Indian Bison** (404) is an imposing animal because of its size and stately build. It is the largest, most massive of all living wild oxen. The original range of its distribution covered the Indian peninsula south of the Himalayas, most of south-eastern Asia, and the Malayan Peninsula, but it has become extinct in many places. It frequents forests, places where there are tall grasses, bamboo jungles and sometimes mountain forests.

Bos gaurus
Bovidae

Body length in old bulls reaches 300—336 cm, the tail measures 80—92 cm, height at the shoulder is 165—190 cm and the weight is 750—1,150 kg. The upturned, extremely massive horns are 60—70 cm long in adult bulls; the maximum length recorded is 80.5 cm. Cows are two-thirds to three-quarters the size of males, and their horns are more slender. Old males have a conspicuous, tall hump on the neck and shoulders. The Gaur gathers in herds of ten to 20 head; old bulls are solitary. Cows are pregnant for 270—285 days and usually bear a single young.

Banteng (405) is one of the most handsome of wild oxen. Its horns are massive but gracefully shaped and adequately long; they differ in form in males and females. The Banteng is smaller then the Arni: it weighs 500—700 kg. Its home is in south-eastern Asia, the Malayan peninsula, Java and Borneo (but, like the Leopard, it is absent from Sumatra). In most of its range, however, it is threatened with extinction. Bantengs live in herds of 20 to 30 head in environments similar to those of the Gaur. Old bulls are often solitary. Gestation lasts 270—280 days; a single young is regularly born.

Bos javanicus
Bovidae

403

Bison or **American 'Buffalo'** (406, 407) is certainly one of the most famous large animals with a rich romantic history. Its destiny — almost complete extinction and subsequent salvation at the eleventh hour — is undoubtedly well-known. In the last century, Bison still lived in North America in herds of hundreds of thousands. They were widespread from the Atlantic to the Pacific coasts, north to 60°N and south to 30°N, withstanding the frost and snow of cruel winters with the same tenacity with which they bore dry, scorching summers. In terms of size, they rank among the largest wild oxen. Old bulls reach 290—310 cm in length; the tail is 55—60 cm, height at the shoulder is 170—180 cm, and the weight is 650—1,000 kg.

Bison bison
Bovidae

404

405

The maximum weight verified is 1,088 kg. Cows are a quarter or a third smaller. Horns of strong bulls are 40—50 cm long; the maximum length is 56.7 cm. Females are pregnant for 270—285 days; a single young is born.

Wisent or **European Bison** (408, 409) in historical times inhabited forests all over Europe, eastwards to the Caucasus and southern Urals. Nowadays, it occurs in the wild only in the Bialowiez Forest on the Polish-Belorussian border, and in a small, restricted location in the Caucasus. It is kept in special parks in many places in Europe. The Wisent is very similar to the American Bison, although it appears to be lighter-built, and the front part of the body in old bulls is less massive. The greatest weight recorded is 952 kg (the usual weight is 680—720 kg); the maximum length of horns is 50.8 cm. The Wisent occurs in two subspecies: the so-called True European Bison (*Bison bonasus bonasus*) (408) which covered most of the

Bison bonasus
Bovidae

406

Heads of the European Wisent (a) and the Caucasian Wisent (b)

407

original range, and the Caucasus Bison (*Bison bonasus caucasicus*) (409), which was somewhat smaller and lived in forests of the Caucasus region. Average gestation lasts 260–270 days; the litter always numbers a single calf.

Yak (410) is the domesticated descendant of the Wild Yak (*Bos mutus*), which has been almost exterminated in its original habitat in the mountainous areas of inner Asia, where it occurred up to altitudes over 6,000 metres. The Yak is protected from unfavourable climatic conditions by its long, hanging hair of a black-brown colour. Old Wild Yak bulls stand 170–175 cm high at the shoulder, and weigh 750–830 kg. The Domestic Yak varies in colour from white to brown and black; it is smaller than the Wild

Bos grunniens
Bovidae

408

409

Yak by up to one-third. Throughout the mountainous regions of central Asia, the Domestic Yak is a universally-used farm animal. Gestation lasts 260—280 days; the litter contains a single, rarely two young.

Aurochs
Bos primigenius
Bovidae

Aurochs was an ancestor of many domesticated cattle breeds. In historical times, it inhabited open forests and their margins, but also steppes over almost all of Europe eastwards to the Urals and the Caucasus, and probably even farther. It

410

The Wild Aurochs used to be the largest European mammal after the Wisent. It became extinct in the early 17th century

411

also lived in Asia Minor and in northern Africa. It disappeared gradually, partly as a result of extermination, but mainly because of the changing conditions in its environment, mostly in connection with the advancing modern age of civilization on the European continent. The Aurochs resembled the Gaur in size (bulls were 260–300 cm long, height at the shoulder was 160–180 cm, and weight was estimated at 700–900 kg, possibly more). Males were deep brown to brown-black, females were brown. As far as we know, the last Aurochs remained in eastern parts of central Europe until the end of the 16th and beginning of the 17th century.

African Buffalo (411)
Syncerus caffer
Bovidae

has become rare or completely exterminated in many places. It originally existed in three forms inhabiting the bush, forests and savanna of the whole of Africa south of the Sahara. The largest form, the Cape Buffalo (*Syncerus caffer caffer*) (411) lives in the bush and savanna from the southern Sudan, Ethiopia and Somalia across eastern Africa to the Cape Province. Old bulls are 270–280 cm long, the tail measures 75–80 cm, and the height at the shoulder is 160–170 cm. They reach 500–750 kg, exceptionally 815 kg, in weight. The characteristically-shaped horns are 70–90 cm long; the maximum recorded length is 127.8 cm. The smallest form, the Forest or Red African Buffalo (*Syncerus caffer nanus*), stands 130–135 cm high at the most; its weight does not exceed 320–330 kg. Gestation in all forms of African buffaloes lasts 300–330 days; usually a single young is born.

412

Chapter 16 WHALES AND DOLPHINS

'And thus . . . among waves whose hand-clappings were suspended by exceeding rapture, Moby Dick moved on, still withholding from sight the full terrors of his submerged trunk, entirely hiding the wretched hideousness of his jaw. But soon the fore part of him slowly rose from the water; for an instant his whole marbleised body formed a high arch, like Virginia's Natural Bridge, and warningly waving his bannered flukes in the air, the grand god revealed himself, sounded, and went out of sight. Hoveringly halting, and dipping on the wing, the white sea-fowls longingly lingered over the agitated pool that he left.'

H. Melville: MOBY DICK
Chapter CXXXII.
The Chase — First Day

Moby Dick, faithfully portrayed in a painting (412) by the Dutch artist R. van Assen, sounded, went out of sight, and the agitated pool that he left slowly vanished. Thus vanished also, a long time ago, the proverbial call 'There she blows!', echoing from the mast-heads of sailing whalers above the waves of seas and oceans. It used to announce an emerging whale, and it has been heard many times from the pages of Melville's immortal novel. A novel bearing not only a profound philosophical message, but in its way constituting a documentary in natural history. Melville's

413

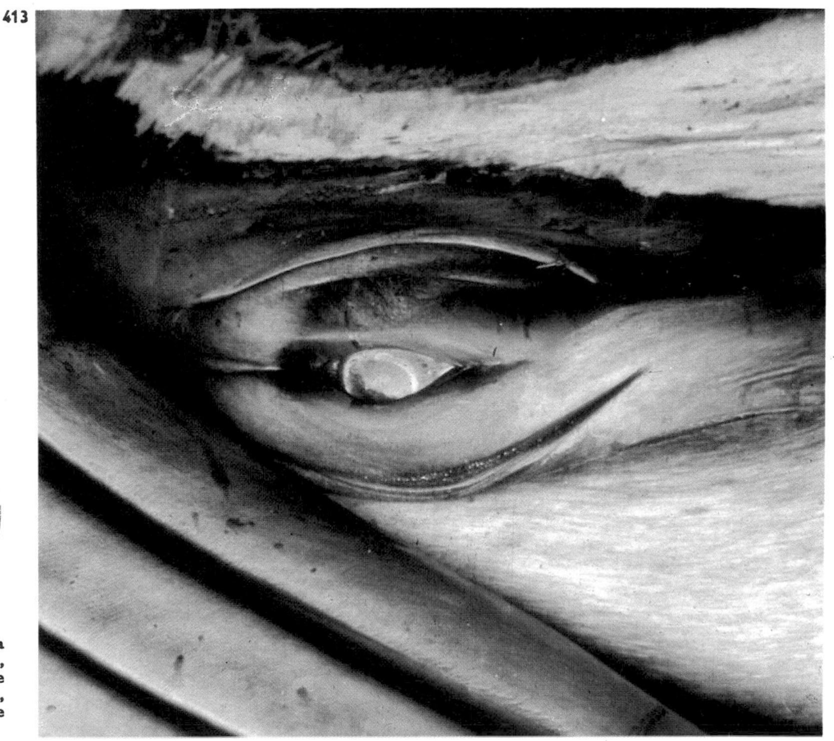

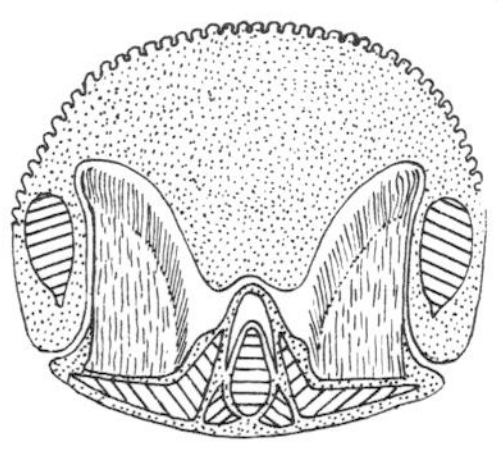

Cross-section of the head of a baleen whale (the Rorqual), showing the location of the whalebone in the mouth, where it functions in the reception of food

414

Moby Dick is justifiably called an 'encyclopaedia of whaling' by many experienced specialists – and that legendary colossal, white sperm whale, which always escaped, really did exist, although it was named by whalers 'Mocha Dick', and it was responsible for a series of shipwrecks and several dozen human lives.

This, however, all took place in the last century, carrying with it the romantic atmosphere of white sails and hand harpoons. Modern times have replaced this romantic spirit (in reality, tough and risky routine), on the one hand by merciless killing of whales with modern harpoons, and on the other hand by an effort to stop these massacres, to protect whales, and to obtain better knowledge of their way of life. By whales, we mean the whole order of cetaceans (Cetacea), in which whales are classified together with other species.

Cetaceans, of which there are about 80 living species, are mammals perfectly adapted to life in the aquatic environment into which they secondarily returned at the beginning of the early Tertiary. In the course of their development, the group divided into two well-distinguished suborders, differing basically by the characteristics from which they take their names: baleen or whalebone whales (Mysticeti) and toothed whales (Odontoceti). The former suborder includes only ten species; the remaining 70 belong to toothed cetaceans.

Most baleen whales are cosmopolitan – they can be found in almost all seas. Only the Bowhead and Grey Whale are confined exclusively to the Northern Hemisphere, while the Pygmy Whale (*Caperea marginata*) of the family Balaenidae, reaching merely 5–6 metres in length, is a typical inhabitant of waters in the Southern Hemisphere. Baleen whales live solitarily, in pairs or in small scattered herds. The teeth are replaced in their jaws by the baleen or whalebone, that is long, horn-like plates, frayed at the margins, which enable whales to catch and filter from sea water large quantities of food: tiny planktonic animals and plants – molluscs, crustaceans, small species of fish, algae, etc. The females (cows) of baleen whales are always somewhat larger than the males (bulls). Gestation lasts 9–10 months (right whales), or 11–12 months (grey whales and rorquals); the female bears a single young.

The three species described next and some others such as the Humpback (*Megaptera novaeangliae*) (423.1) and the Sei Whale (*Balaenoptera borealis*) (423.2), rank among baleen whales, in the rorqual family (Balaenopteridae). The second family of baleen whales, the grey whales (Eschrichtidae), is represented by a sole member, the Grey Whale (*Eschrichtius robustus*) (423.5). The third family, the right whales (Balaenidae), includes – as well as the above-mentioned Pygmy Whale – the

Bowhead Right Whale (*Balaena mysticetus*) (423.7), and the Black Right Whale (*Eubalaena glacialis*) (423.6), both reaching a length of 16–18 metres and a weight of 50,000–60,000 kg. Rorquals can be distinguished on sight from right whales by the several dozen longitudinal grooves on the lower front part of their body, and by the presence of a small dorsal fin, which right whales lack (the only exception being the Pygmy Whale).

Blue Whale or **Giant Rorqual** (423.4) is the largest cetacean and the largest mammal. It attains a length of up to 30 metres, exceptionally 33 metres, and its weight can reach 163,000 kg (the greatest weight so far recorded; the usual weight is 100,000–120,000 kg).
Balaenoptera musculus
Balaenopteridae

Fin Whale or **Rorqual** (413 – detail of mouth and eye, 423.3) is the biggest whale next to the Blue Whale. It reaches 25 metres in length and weighs about 80,000 kg.
Balaenoptera physalus
Balaenopteridae

Minke Whale or **Lesser Rorqual** (414) reaches only 8–9 metres in length, and does not weigh more than 6,000–8,000 kg.
Balaenoptera acutorostrata
Balaenopteridae

Toothed whales (Odontoceti) are distinguished from baleen whales above all by having teeth, not whalebone: they can trap, grasp and process food of larger dimensions. They feed on fish, even on large species, exceptionally on other cetaceans and pinnipeds, larger cephalopods, crustaceans, etc. Toothed whales have only one functional respiratory orifice, while baleen whales have two. The lower jaw in toothed whales is fused in front, forming the symphysis, and the skull is asymmetrical, both in the shape of individual bones, and in their dimensions. In baleen whales,

415

416

the lower jaw does not fuse in front and the skull is always more or less symmetrical. The suborders also differ in the structure of internal organs, for example, in the digestive system: the stomachs are differently constructed, and toothed whales have in addition a vermiform appendix. Further differences exist in the nature of the sounds emitted.

Relatively few species of toothed whale are widespread in all seas; their range of distribution is usually limited to a specific geographical area, however, vast as these may be. Unlike baleen whales, the males of toothed whales are often conspicuously larger than females. Gestation in most toothed whales takes 9–12 months, and longer in a few species: 14 months in the Beluga and 16 months in the Sperm Whale. Litters number a single young, as in baleen whales.

The suborder of toothed whales is divided into several families: freshwater dolphins (Susuidae), belugas and narwhals (Monodontidae), dolphins and porpoises (Delphinidae), beaked whales (Hyperoodontidae), and sperm whales (Physeteridae).

Inia or **Amazon Dolphin** (415, 416), representing the family of freshwater or river dolphins, is a cetacean 2–2.5 metres long, one of the few inhabiting fresh waters. It occurs in river systems of the Amazon and Orinoco.
Inia geoffrensis
Susuidae

Beluga or **White Whale** (417) is pure white when adult and reaches 5–6 metres in length, exceptionally even more (the maximum length recorded is 6.67 metres). Its weight is 1,000–1,500 kg, exceptionally 1,850 kg. The Beluga has a circumpolar distribution in cold waters of the Northern Hemisphere. It sometimes enters the estuaries of large rivers.
Delphinapterus leucas
Monodontidae

Common Dolphin (418) is the most typical member of the dolphin family, the most numerous cetacean family. It includes almost 50 species, from small ones 120 to 150 cm in length to large species, 8—9 metres long. The Common Dolphin ranks among the smaller ones: it is 2—2.6 metres long, and it weighs 100—180 kg. Except in the coldest waters, it is widespread in all oceans and seas, including the Black Sea.

Delphinus delphis
Delphinidae

Striped Dolphin (419) reaches a length of about 2.25 metres, and weight of 100—110 kg. The species is restricted to the North Pacific Ocean, in waters of the cold and temperate zones.

Lagenorhynchus obliquidens
Delphinidae

417

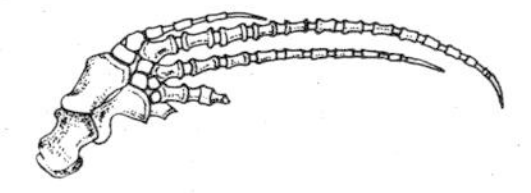

Skeleton of front flipper-like limbs in cetaceans demonstrates different degrees of reduction of the toes, and specialization for life in an aquatic environment. Above: A baleen whale (the Bowhead); below: a toothed cetacean (the Pilot Whale, ***Globicephala melaena*****)**

418

419

Bottle-nosed Dolphin (420) is a dark-coloured, medium-sized dolphin, attaining up to 3–3.4 metres in length, and a weight of 250–450 kg. Its distribution corresponds largely to that of the Common Dolphin. It has become one of the best known cetaceans, since it is one of the most common inhabitants of dolphinaria and oceanaria. The species has been the subject of many investigations which have proved the now generally recognized superior intelligence of cetaceans, comparable only to the intelligence of highest primates (naturally with the exception of Man).
Tursiops truncatus
Delphinidae

Orca or **Killer Whale** (421) is the largest dolphin. The males grow to 8–9 metres in length, and their weight exceeds 4,500 kg (allegedly almost 8,000 kg). The females are somewhat smaller. The name 'Killer Whale' was given to these cetaceans by whalers: it is exaggerated, for the Orca does nothing that other animals would not do — it simply looks after its stomach. The disreputable epithet was probably a result of the nature of the Orca's victims, which are more obvious than those of other cetaceans: large fish, seals, sea lions and other cetaceans. Stories of groups of orcas killing and tearing to pieces even some large whales, are unlikely to be true. They must have been triggered off by the fact that orcas sometimes rip pieces of flesh from harpooned, but always dead whales. The Orca is cosmopolitan: it is found in both cold and warm waters all over the world.
Orcinus orca
Delphinidae

Sperm Whale (412, 422, 423.8) was partly introduced as 'Moby Dick' at the beginning of this chapter. Although normally dark-coloured, sperm whales can exceptionally be pure or partly white. White was the colour of Moby Dick and of
Physeter macrocephalus
Physeteridae

420

its living model, Mocha Dick. Old bulls can reach 23 or even 24 metres in length, but such large specimens are no longer found 'thanks' to whaling massacres. At present, Sperm Whales 18–19 metres long are regarded as rarities. The weight of old males could reach 50,000 to 80,000 kg (again, these giants of 80 tons are no longer encountered). The females are much smaller: 12–14 metres at the most. The gigantic, boulder-shaped head with a blunt front occupies up to two-fifths of the total length of old males. It was this head, that formidable power-hammer, which used to break the rigging of sailing whalers. The enormous teeth, situated in the extremely narrow lower jaw, can be 33 cm long including the root; their weight can reach 1,900 grams. The Sperm Whale has the largest brain of all animals: 8,340–9,200 grams (man's brain averages 1,400 grams). In addition to these and many other peculiarities, the Sperm Whale is known for two substances produced in its body. The first is spermaceti, filling the huge cavity of the colossal head in a quantity of 5,000–6,000 kg; it is used in cosmetics and pharmacology. The second substance is ambergris; this is not found in all sperm whales, but only in the intestines of certain individuals. It is also found floating at sea, eliminated from the digestive system with the excreta. It is a waxy, usually dark-coloured substance, highly valued in the manufacture of the most costly perfumes. At the beginning of 1978 its market price was

421

422

approximately £316 for 1 kg! The largest piece of ambergris, discovered in a sperm whale killed in 1912, weighed 454.5 kg; the usual weight of pieces of this precious substance is, however, merely several hundred grams or several kilograms at the most. The Sperm Whale is resident in all seas and oceans, chiefly in tropical and temperate zones, and is solitary, especially old males, or lives in herds of up to several dozen individuals. Its diet consists of large fish, but mainly of giant squids, for example of the genus *Architeuthis* (423.9). The Sperm Whale is capable of diving 1,000–2,000 metres in pursuit of its prey.

The limited space of this chapter prevents discussion of other interesting details of the life of cetaceans, particularly their method of echolocation. Let us not forget, however, that cetaceans are endangered by the ominous and ruthless exploitation of Nature which accompanies so-called 'advanced' human civilization. Many species other than cetaceans are threatened with total extinction. It is the duty of everyone to strive — each according to his ability — towards better understanding of Nature and her riches. Although it was not our main purpose, we have at least tried in this book to suggest that the animals now threatened by extermination or extinction are many — too many . . .

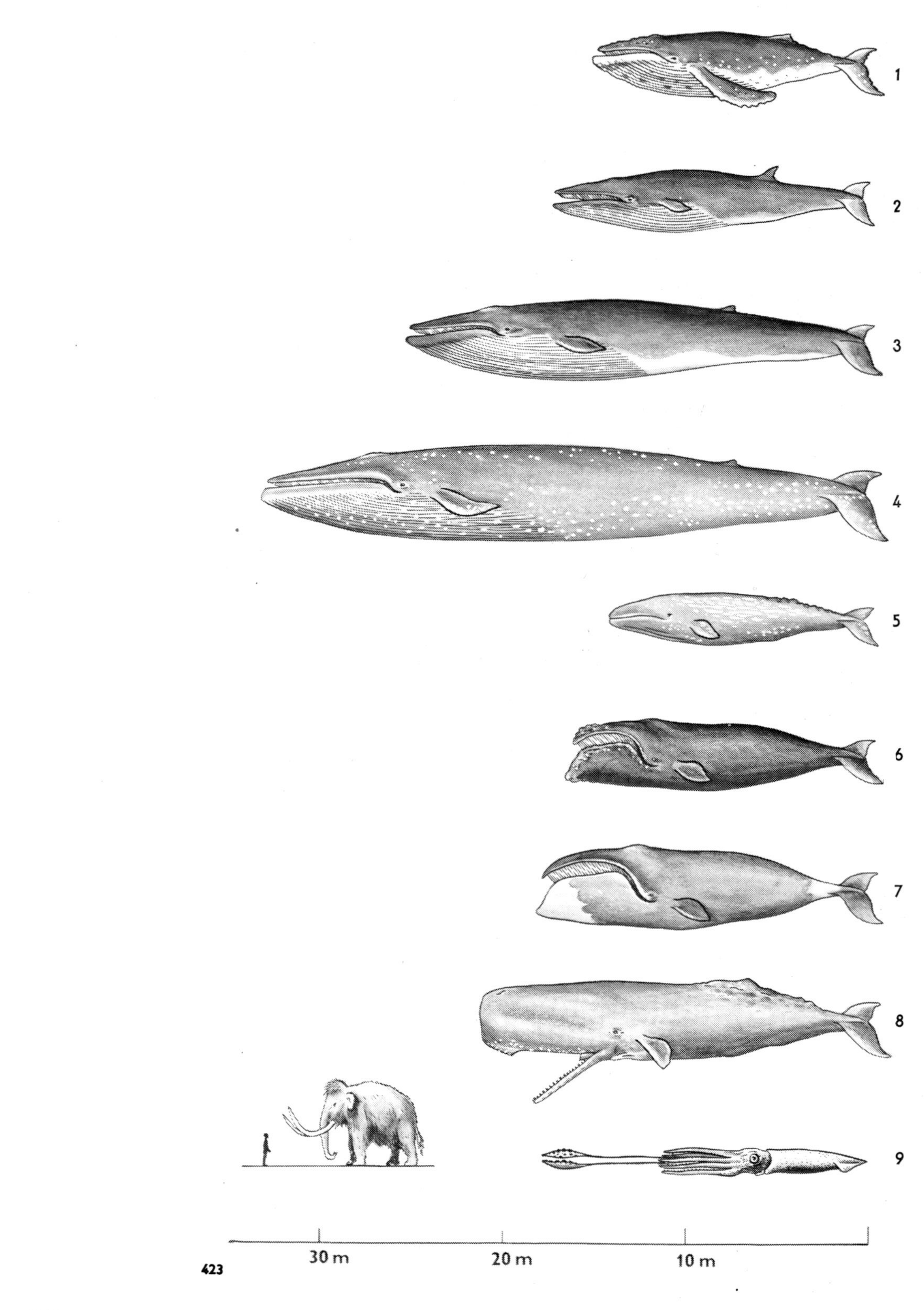

423

SUGGESTIONS FOR FURTHER READING

Bourlière F.: The Natural History of Mammals, Harrap, London, 1955

Burt W. H. and Grossenheinder R. P.: A Field Guide to the Mammals, Haughton Mifflin Co., Boston, 1964

Corbet G. B.: The Terrestrial Mammals of Western Europe, Foulis, London, 1966

Corbet G. B. and Southern H. N. (Eds)*:* The Handbook of British Mammals, Blackwell, Oxford, 1977

Dorst J. and Dandelot P.: Field Guide to the Larger Mammals of East Africa, Collins, London, 1970

DuPlaix N. and Simon N.: World Guide to Mammals, Crown, New York, 1976; Octopus, London, 1977

Ewer R. F.: The Ethology of Mammals, Logos, New York, 1968

Ewer R. F.: The Carnivores, Weidenfield & Nicolson, London, 1973

Fisher J. (and others): The Red Book: Wild Life in Danger, Collins; London, 1969

Groves C. P.: Horses, Asses and Zebras, David & Charles, Newton Abbot, 1974

Grzimek B.: Animal Life Encyclopaedia, Vols 10–13 (Mammals I–IV), Von Nostrand Reinhold, New York, 1972

Hanák V.: A Colour Guide to Familiar Mammals, Octopus, London, 1977

Hanney P. W.: Rodents, David & Charles, Newton Abbot, 1975

Harris C. J.: Otters, Weidenfield & Nicolson, London, 1968

Harrison R. J. and King J.: Marine Mammals, Hutchinson, London, 1965

Harrison Mathews L.: The Life of Mammals, Weidenfield & Nicolson, London, 1969 and 1971

Leen N. and Novick A.: The World of Bats, Edita, Lausanne, 1969.

Napier J. R. and Napier P. H.: Handbook of Living Primates, Academic Press, London and New York, 1967

Ride W. D. L.: A Guide to the Native Mammals of Australia, Oxford University Press, Melbourne and London, 1970

Sanderson I. T.: Living Mammals of the World, Hamish Hamilton, London; Chanticleer, New York, 1955

Schultz A.: The Life of Primates, Weidenfield & Nicolson, London, 1969

Thompson H. V. and Worden A. N.: The Rabbit, Collins, London, 1956

Tyndale-Byscoe H.: The Life of Marsupials, Arnold, London, 1973

Van den Brink F. H.: A Field Guide to the Mammals of Britain and Europe, Collins, London, 1976

Walker E. P.: Mammals of the World, John Hopkins, Baltimore, 1968

INDEX OF COMMON NAMES

(Bold figures refer to numbers of illustrations)

Aardvark 123, **117**
Aardwolf 187, **211**
Addax 302, **360**
Adjag 196
Alpaca 272
Anteater, Great 119, **109, 110**
Anteater, Marsupial 38, **16**
Anteater, Two-toed 118
Antelope, Harnessed 294, **348**
Antelope, Indian Four-horned 289
Antelope, Pronghorn 289, **343**
Antelope, Roan 300, **357**
Antelope, Sable 301, **358**
Aoudad 319, **394**
Apara 122, **114**
Ape, Barbary 103
Argali 323, **398**
Argali, Altai 323, **398**
Argali, Marco Polo 323
Armadillo, Hairy 121, **113**
Armadillo, Nine-banded 122, **115**
Arni 327, **403**
Ass, African Wild 234, **284**
Ass, Asiatic Wild 235, **285**
Ass, Central Asiatic Wild 235
Ass, Somalian 234
Aurochs 332
Aye-Aye 82, **69**

Babirussa 265, **313**
Baboon, Anubis 104, **96**
Baboon, Gelada 103, **95**
Badger, American 165, **181**
Badger, Honey 163, **179**
Badger, Old World 164, **180**
Bandicoot, Rabbit 41
Banteng 328, **405**
Bat, African Hammerhead Fruit 69
Bat, Common Noctule 76, **59**
Bat, European Free-tailed 77
Bat, Free-tailed 77, **63**
Bat, Greater Horseshoe 70, **53, 54**
Bat, Lesser Horseshoe 71, **55**
Bat, Long-eared 76, 58
Bat, Long-winged 77, **61, 62**
Bat, Mexican Free-tailed 77
Bat, Mouse-eared 77, **60**
Bat, Short-nosed Fruit 68, **50**
Bat, Straw-coloured 68, **52**
Bat, Vampire 73, **57**
Bat, Wahlberg's Epauletted Fruit 68, **51**
Bear, American Black 177, **196**
Bear, Asiatic Black 177, **197**
Bear, Brown 173, **193–195**
Bear, European Brown 174, **193**
Bear, Kodiak 176, **195**
Bear, Malayan Sun 179, **199**
Bear, Polar 181, **201**
Bear, Sloth 177, **198**
Bear, Spectacled 179, **200**
Beaver, European 132, **131, 132**
Beluga 338, **417**
Binturong 182, **205**
Bison 329, **406, 407**
Bison, Caucasus 331, **409**
Bison, European 330, **408, 409**
Bison, Indian 328, **404**
Bison, True European 330, **408**
Blackbuck 313, **384**
Blesbok 303, **363**
Bluebuck 297, 301
Boar, Wild 263, **308**
Bobcat 203, **242**
Bongo 299, **356**
Bontebok 303, **363**
Buffalo, African 333, **411**
Buffalo, American 329, **406, 407**
Buffalo, Cape 333, **411**
Buffalo, Forest 333
Buffalo, Red African 333
Buffalo, Water 327, **402**
Buffalo, Wild Water 327, **403**
Bull, Blue 301, **359**
Burunduk 131, **127**
Bush Baby, Senegal 83, **72**
Bushbuck 294, **348**

Cacomistle 167, **185**
Camel, Arabian 269, **318**
Camel, Bactrian 268, **317**
Camel, Wild 269
Capuchin, Brown 94, **84**
Capybara 151, **161**
Caracal 203, **243**
Cat, African Desert 200, **238**
Cat, African Golden 204, **244**
Cat, African Tiger 204, **244**
Cat, Amur Forest 198
Cat, European Wild 199, **235, 236**
Cat, Flat-headed 198, **229**
Cat, Indian Desert 199, **236**
Cat, Indian Golden 205, **245**
Cat, Jungle 200, **237**
Cat, Leopard 198, **230**
Cat, Marbled 205, **246**
Cat, Native 33, **10**
Cat, Ocelot 198, **231**
Cat, Pallas' 200, **239**
Cat, Pampas 199, **233**
Chamois 315, **386**
Cheetah 215, **258**
Chevrotain 273, **321**
Chimpanzee 115, **108**
Chinchilla 154, **164**
Chipmunk, Eastern 132, **128**
Chital 280, **331**
Civet, African 181, **202**
Civet, Palm 182, **204**
Coatimundi, Red 171, **189**
Coatimundi, White-lipped 171, **190**
Colobus, Black-and-White 96, **87**
Cuscus, Spotted 42, **22**

Deer, Axis 280, **331**
Deer, Barasinga 283, **334**
Deer, Barking 275, **324**
Deer, Brocket 276, **325**
Deer, Chinese Water 274, **323**
Deer, Fallow 283, **335**
Deer, Hog 280, **332**
Deer, Indian Mouse 273, **321**
Deer, Javan Mouse 273, **322**
Deer, Mule 276, **326**
Deer, Musk 275
Deer, Père David's 279, **330**
Deer, Red 285, **337, 338**
Deer, Roe 285, **339**
Deer, White 285, **338**
Degu 155, **166**
Desman, Pyrenean 58
Desman, Russian 57, **40**
Dhole 196, **227**
Dikdik, Kirk's 291, **346**
Dingo 191, **215**
Dog, Africain Wild 197, **228**
Dog, Black-tailed Prairie 130, **126**

Dog, Bush 196, **225**
Dog, Raccoon 195, **223**
Dolphin, Amazon 338, **415, 416**
Dolphin, Bottle-nosed 341, **420**
Dolphin, Common 339, **418**
Dolphin, Striped 339, **419**
Dormouse, Common 147, **155**
Dormouse, Fat 146, **153**
Dormouse, Garden 146, **154**
Doroucouli 89, **79**
Drill 108, **99**
Dromedary 269, **318**
Dugong, Indian 244, **292**
Duiker, Black-fronted 290, **344**
Duiker, Yellow-backed 290, **345**
Dunnart, Fat-tailed 33, **11**
Dzheyran 310, **379**

Echidna, Australian 25, **1, 2**
Echidna, Bruijn's 27, **3, 4**
Eland, Cape 298, **354**
Eland, Giant 298, **355**
Elephant, African 251, **297, 298**
Elephant, Bush 253, **297, 298 left**
Elephant, Forest 253, **298 right**
Elephant, Indian 253, **299, 300, 301**
Elk 285, **337, 338**
Ermine 158, **169, 170**
Eyra 199, **234**

Fanaloka 187, **210**
Fennec 195, **222**
Fosa 187
Fox, African Big-eared 196, **226**
Fox, Arctic 193, **220**
Fox, Blue 194
Fox, Common 192, **219**
Fox, Korsak 194, **221**
Fox, Red 192, **219**

Gaur 328, **404**
Gazelle, Giraffe 310, **376, 377**
Gazelle, Grant's 311, **380**
Gazelle, Persian 310, **379**
Gazelle, Thomson's 310, **378**
Gemsbok 302, **362**
Genet, Common 182, **203**
Gerenuk 310, **376, 377**
Gibbon, White-handed 112, **104**
Giraffe 286, **341, 342**
Giraffe, Reticulated 287, **341**
Giraffe, Rothschild's 288, **342**
Glider, Pygmy 45, **25**
Glider, Sugar 44, **24**
Gnu, Brindled 305, **368, 369**
Gnu, White-bearded 305, **369**
Gnu, White-tailed 305, **370**
Goat, Rocky Mountain 315, **387**
Goat, Wild 316, **388**
Goral 315, **385**
Gorilla 114, **107**
Gorilla, Lowland 114
Gorilla, Mountain 114
Grison 162, **177**
Grizzly, North American 176, **194**
Guanaco 271, **319**
Guenon, Red 110, **103**

Hamster, Common 138, **137**
Hamster, Crested 139, **141**
Hamster, Golden 139, **139**
Hamster, Roborowsky's Dwarf 139, **140**
Hare, Blue 258, **305**
Hare, Cape Spring 137, **135**
Hare, Daurian Piping 257, 303
Hare, European 257, **304**
Hare, Snowshoe 259, **306**
Hartebeest, Coke's 305, **366**
Hartebeest, Hunter's 304, **364**
Hartebeest, Lichtenstein's 305, **367**
Hartebeest, Red 304, **366, 367**
Hedgehog, African 61, **44**
Hedgehog, European 60, **43**
Hippopotamus 267, **316**
Hippopotamus, Pygmy 265, **315**
Hog, Giant Forest 263, **312**
Hog, Wart 263, **311**
Horse, Przewalski's 233, **282, 283**
Humpback 336, **423.1**
Hyaena, Brown 190, **214**
Hyaena, Spotted 188, **212**
Hyaena, Striped 190, **213**
Hyrax, Johnston's 248
Hyrax, Rock 248, **294**
Hyrax, Tree 248, **296**
Hyrax, Yellow-spotted 248, **295**

Ibex, Alpine 317, **389**
Ibex, Caucasus 318, **391**
Ibex, Dagestan 318, **392**
Ibex, Ethiopian 319
Ibex, Mountain 316, **389**
Ibex, Nubian 317
Ibex, Siberian 317, **390**
Impala 309, **375**
Indri 81
Inia 338, **415, 416**

Jackal, Black-backed 192, **218**
Jackal, Common 192, **217**
Jaguar 208, **251**
Jaguarondi 199, **234**
Jerboa, Siberian 147, **156**
Jird, Southern 142, **146**

Kangaroo, Black-faced 48, **30**
Kangaroo, Red 47, **29**
Kangaroo, Rufous Rat 46, **26**
Kangaroo, Tree 50, **33**
Kangaroo, Western Grey 48, **30**
Kerabau 327, **402**
Kiang 237, **286**
Kinkajou 168, **186**
Klipspringer 291, **347**
Koala 40, **18, 19**
Kob 309, **373**
Konzi 305, **367**
Kudu, Greater 295, **352, 353**
Kudu, Lesser 295, **351**
Kulan 235

Langur, Capped 97, **88**
Langur, Douc 97, **92**
Lechwe 309, **374**
Lemur, Mongoose 81, **67**
Lemur, Ring-taled 80, **66**
Lemur, Ruffed 81, **68**
Leopard 206, **167, 249, 250**
Leopard, Clouded 206, **248**
Leopard, Snow 215, **257**
Lion 208, **252, 253, 254**
Lion, Cape 209
Lion, Mountain 205, **247**
Llama, Domesticated 272
Loris, Slender 83, **70**
Loris, Slow 83, **71**
Lynx, Bay 203, **242**
Lynx, Desert 203, **243**
Lynx, Northern 203, **241**

Macaque, Crab-eating 103
Macaque, Japanese 103, **94**
Macaque, Lion-tailed 103
Macaque, Rhesus 100, **93**
Manatee, Wide-nosed 244, **293**
Mandrill 107, **97, 98**

Mangabey, Sooty 108, **100**
Mara 152, **162**
Margay 199, **232**
Markhor 318, **393**
Marmoset, Golden Lion 88, **77**
Marmoset, Silvery 89, **78**
Marmot, Alpine 128, **123**
Marmot, Bobac 129, **124**
Marten, American Pine 159, **173**
Marten, Beech 160, **175**
Marten, Kharza 162, **176**
Marten, Pine 160, **174**
Mole, American Star-nosed 56
Mole, Common 56, **39**
Mole, Grant's Desert 58, **41**
Mole, Marsupial 39, **17**
Mongoose, African 184, **206**
Mongoose, Banded 187, **208**
Mongoose, Common 184, **207**
Monkey, Black Spider 94, **85**
Monkey, Common Squirrel 95, **86**
Monkey, De Brazza's 109, **101**
Monkey, Diana 110, **102**
Monkey, Dusky Leaf 97, **89**
Monkey, Proboscis 97, **90, 91**
Monkey, Red Howler 94, **83**
Moose 276, **328**
Moose, Alaska 277, **328**
Mouflon 319, **395, 396**
Mouse, Deer 137, **136**
Mouse, Egyptian Spiny 145, **151**
Mouse, House 145, **149**
Mouse, Marsupial 34, **12**
Mouse, Northern Birch 148, **157**
Mouse, Striped Grass 146, **152**
Mouse, Yellow-necked Field 145, **150**
Muntjac 275, **324**
Muskrat 141, **144**

Newara 184, **207**
Nilgai 301, **359**
Numbat 38, **16**
Nutria 154, **165**
Nyala, Lowland 294, **350**

Ocelot, Tree 199, **232**
Ocelot, Tree 199, **232**
Okapi 286, **340**
Opossum, Mouse 31, **8**
Opossum, North American 32, **9**
Orang-utan 112, **64, 106**
Orca 341, **421**
Oryx, Beisa 302, **362**
Oryx, Scimitar-horned 302, **361**
Otter, Common 165, **183**
Otter, Sea 166, **184**
Ounce 215, **257**
Ox, Musk 327, **401**

Panda, Giant 171, **192**
Panda, Lesser 171, **191**
Pangolin, Chinese 122, **116**
Peccary, Collared 265, **314**
Phalanger, Brush-tailed 42, **21**
Phalanger, Honey 43
Phalanger, Striped 43, **23**
Pig, Bush 263, **309, 310**
Pig, South African Bush 263, **310**
Pig, West African Bush 263, **309**
Pika, American 257, **302**
Platypus, Duck-billed 28, **5, 6**
Polecat, Common 159, **171**
Polecat, Eversmann's 159, **172**
Polecat, Steppe 159, **172**
Porcupine, North American 149, **160**
Porcupine, White-tailed 148, **158**
Potto 84, **73**
Pudu 276, **327**
Puma 205, **247**

Quagga 240, **241**

Rabbit, European Wild 259, **307**
Raccoon, Crab-eating 170, **188**
Raccoon, Ring-tailed 169, **187**
Rat, Black 143, **148**
Rat, Florida Wood 139, **138**
Rat, Lesser Mole 142, **147**
Rat, Merriam's Kangaroo 135, **133**
Rat, Naked Mole 149, **159**
Rat, Norwegian 144
Reedbuck, Bohor 308, **371**
Reindeer 278, **329**
Rhinoceros, Black 229, **275**
Rhinoceros, Indian One-horned 227, **272, 273**
Rhinoceros, Square-lipped 230, **276, 277**
Rhinoceros, Sumatran 229, **274**
Rhinoceros, White 230, **276, 277**
Ringtail 167, **185**
Rorqual 337, **413, 423.3**
Rorqual, Giant 337, **423.4**
Rorqual, Lesser 337, **414**

Sable, North Asiatic 159
Saiga 312, **383**
Saki, Hairy 93, **82**
Saki, Pale-headed 91, **81**
Sambar 281, **333**
Sassaby 304, **365**
Sea Cow, Steller's 243, 244
Sea Lion, Australian 218, **259**
Sea Lion, Californian 219, **260**
Sea Lion, Steller's 219, **261**
Seal, Baïkal Ringed 224, **266**
Seal, Grey 224, **268**
Seal, Harbour 221, **264, 265**
Seal, Hawaiian 224, **267**
Seal, Hooded 225, **269**
Seal, South Atlantic Elephant 225, **270, 271**
Serigala 196
Serval 201, **240**
Sheep, Alaskan 323, **397**
Sheep, Barbary 319, **394**
Sheep, Bighorn 323, **397**
Sheep, Central Asiatic Wild 323, **398**
Sheep, Kamchatka 323
Shrew, Alpine 62, **35**
Shrew, Bicolour White-toothed 64, **48**
Shrew, Common Tree 79, **65**
Shrew, Eurasian Water 63, **46**
Shrew, House 64
Shrew, Lesser White-toothed 63, **47**
Shrew, North African Elephant 59, **42**
Shrew, Otter 54, **37**
Shrew, Piebald 65, **49**
Shrew, Pygmy 62, **45**
Shrew- Savi's Pygmy 61
Siamang 112, **105**
Sika 285, **336**
Sitatunga 294, **349**
Skunk, Common 165, **182**
Skunk, Striped 165, **182**
Sloth, Two-toed 120, **112**
Solenodon, Cuban 53, **36**
Souslik, European 126, **122**
Springbok 311, **381, 382**
Squirrel, African Ground 132, **130**
Squirrel, Arctic Ground 126, **120**
Squirrel, California Ground 126, **121**

Squirrel, Grey 126, **119**
Squirrel, Pel's Scaly-tailed 137, **134**
Squirrel, Red 125, **118**
Squirrel, Southern Flying 132, **129**
Stoat 158, **169, 170**
Suricata 187, **209**

Tahr 325, **399**
Tahr, Himalayan 325, **399**
Takin 326, **400**
Tamandua 120, **111**
Tamarin, Cotton-head 87, **75**
Tamarin, Emperor 88, **76**
Tapir, Brazilian 231, **278**
Tapir, Central American 231
Tapir, Malayan 233, **280, 281**
Tapir, Mountain 232, **279**
Tarsier, Philippine 84, **74**
Tasmanian Devil 34, **13, 14**
Tenrec 55, **38**
Tenrec, Tail-less 55
Tiger 211, **255, 256**
Tiger, American 208, **251**
Tiger, Indian 213, **256**
Tiger, Ussurian 213, **255**
Topi, Swift 304, **365**
Tur, Caucasus 318, **391**
Tur, Dagestan 318, **392**

Uakari, Bald 90, **80**
Ucumari 179, **200**

Vampire, Heart-nosed False 72, **56**
Vicuña 272, **320**
Viscacha 153, **163**
Vole, Bank 141, **145**
Vole, Common 140, **143**
Vole, Water 139, **142**

Wallaby, New Guinea Mountain 50, **34**
Wallaby, Parry's 50, **32**
Wallaby, Ring-tailed Rock 47, **28**
Wallaby, Rock 46, **27**
Wallaby, Swamp 48, **31**
Walrus 220, **262, 263**
Wapiti 285, **337, 338**
Waterbuck 308, **372**
Waterbuck, Defassa 308, **372**
Waterbuck, Kringgat 308
Weasel, Common 157, **168**
Whale, Black Right 337, **423.6**
Whale, Blue 337, **423.4**
Whale, Bowhead Right 337, **423.7**
Whale, Fin 337, **413, 423.3**
Whale, Grey 336, **423.5**
Whale, Killer 341, **421**
Whale, Minke 337, **414**
Whale, Pilot 339
Whale, Pygmy 336
Whale, Sei 336, **423.2**
Whale, Sperm 341, **412, 422, 423.8**
Whale, White 338, **417**
Wildebeest 305, **368, 369**
Wildebeest, Blue 305, **368**
Wisent 330, **408, 409**
Wolf 192, **216**
Wolf, Asiatic Red 196, **227**
Wolf, Maned 195, **224**
Wolf, Tasmanian 35, **15**
Wolverine 162, **178**
Wombat, Hairy-nosed 41, **20**
Woodchuck 130, **125**

Yak 331, **410**
Yak, Wild 331

Zebra, Böhm's Steppe 241, **291**
Zebra, Cape Mountain 237
Zebra, Chapman's Steppe 241, **289, 290**
Zebra, Grevy's 237, **287**
Zebra, Hartmann's Mountain 237, **288**
Zebra, Mountain 237, **288**
Zebra, Steppe 241, **289, 290, 291**

INDEX OF SCIENTIFIC NAMES

(Bold figures refer to numbers of illustrations)

Acinonyx jubatus 215, **258**
Acomys cahirinus 145, **151**
Acrobates pygmaeus 45, **25**
Addax nasomaculatus 302, **360**
Aepyceros melampus 309, **375**
Aeprymnus rufescens 46, **26**
Ailuropoda melanoleuca 171, **192**
Ailurus fulgens 171, **191**
Alcelaphus buselaphus 304, **366, 367**
Alcelaphus buselaphus cokii 305, **366**
Alcelaphus buselaphus lichtensteini 305, **367**
Alces alces 276, **328**
Alces alces gigas 277, **328**
Allactaga sibirica 147, **156**
Alopex corsac 194, **221**
Alopex lagopus 193, **220**
Alouatta seniculus 94, **83**
Ammotragus lervia 319, **394**
Anomalurus peli 137, **134**
Antechinus bellus 34, **12**
Antidorcas marsupialis 311, **381, 382**
Antilocapra americana 289, **343**
Antilope cervicapra 313, **384**
Aotus trivirgatus 89, **79**
Apodemus flavicollis 145, **150**
Architeuthis 343, **423.9**
Arctictis binturong 182, **205**
Arctocephalus doriferus 218, **259**
Arvicola terrestris 139, **142**
Ateles paniscus 94, **85**
Axis axis 280, **331**

Babyrousa babyrussa 265, **313**
Balaena mysticetus 337, **423.7**
Balaenoptera acutorostrata 337, **414**
Balaenoptera borealis 336, **423.2**
Balaenoptera musculus 337, **423.4**
Balaenoptera physalus 337, **413, 423.3**
Bassariscus astutus 167, **185**
Bison bison 329, **406, 407**
Bison bonasus 330, **408, 409**
Bison bonasus bonasus 330, **408**
Bison bonasus caucasicus 331, **409**
Boocercus euryceros 299, **356**
Bos gaurus 328, **404**

Bos grunniens 331, **410**
Bos javanicus 328, **405**
Bos mutus 331
Bos primigenius 332
Boselaphus tragocamelus 301, **359**
Bradypus didactylus 244
Bubalus arnee 327, **403**
Bubalus bubalis 327, **402**
Budorcas taxicolor 326, **400**

Cacajao calvus 90, **80**
Callithrix argentata 89, **78**
Camelus bactrianus 268, **317**
Camelus dromedarius 269, **318**
Camelus ferus 269
Canis aureus 192, **217**
Canis dingo 191, **215**
Canis hallstromi 191
Canis lupus 192, **216**
Canis mesomelas 192, **218**
Caperea marginata 336
Capra aegagrus 316, **388**
Capra caucasica 318, **391**
Capra cylindricornis 318, **392**
Capra falconeri 318, **393**
Capra ibex 316, **389**
Capra ibex ibex 317, **389**
Capra ibex nubiana 317
Capra sibirica 317, **390**
Capreolus capreolus 285, **339**
Caracal caracal 203, **243**
Castor fiber 132, **131, 132**
Cebus apella 94, **84**
Cephalophus nigrifrons 290, **344**
Cephalophus sylvicultor 290, **345**
Ceratotherium simum 230, **276, 277**
Cercocebus torquatus 108, **100**
Cercopithecus diana 110, **102**
Cercopithecus neglectus 109, **101**
Cervus elaphus 285, **337, 338**
Cervus nippon 285, **336**
Chinchilla laniger 154, **164**
Choeropsis liberiensis 265, **315**
Choloepus didactylus 120, **112**
Choloepus hoffmanni 244
Chrysocyon brachyurus 195, **224**
Citellus beechei 126, **121**
Citellus citellus 126, **122**
Citellus parryi 126, **120**
Clethrionomys glareolus 141, **145**
Colobus polycomos 96, **87**

Condylura cristata 56
Connochaetes gnu 305, **370**
Connochaetes taurinus 305, **368, 369**
Connochaetes taurinus albojubatus 305, **369**
Connochaetes taurinus taurinus 305, **368**
Cricetus cricetus 138, **137**
Crocidura leucodon 64, **48**
Crocidura suaveolens 63, **47**
Crocuta crocuta 188, **212**
Cryptoprocta ferox 187
Cuon alpinus 196, **227**
Cyclopes didactylus 118
Cynomys ludovicianus 130, **126**
Cynopterus brachyotis 68, **50**
Cystophora cristata 225, **269**

Dactylopsila trivirgata 43, **23**
Dama dama 283, **335**
Damaliscus dorcas 303, **363**
Damaliscus hunteri 304, **364**
Damaliscus lunatus 304, **365**
Damaliscus lunatus korrigum **304**
Dasypus novemcinctus 122, **115**
Dasyurus quoll 33, **10**
Daubentonia madagascarensis 82, **69**
Delphinapterus leucas 338, **417**
Delphinus delphis 339, **418**
Dendrohyrax dorsalis 248, **296**
Dendrolagus matschiei 50, **33**
Desmana moschata 57, **40**
Desmodus rotundus 73, **57**
Diceros bicornis 229, **275**
Didelphis marsupialis 32, **9**
Didermoceros sumatrensis 229, **274**
Diplomesodon pulchellum 65, **49**
Dipodomys merriami 135, **133**
Dolichotis patagona 152, **162**
Dorcopsis macleayi 50, **34**
Dugong dugong 244, **292**

Echinops telfairi 55, **38**
Eidolon helvum 68, **52**
Elaphurus davidianus 279, **330**
Elephantulus rozeti 59, **42**
Elephas maximus 253, **299, 300, 301**
Eliomys quercinus 146, **154**
Enhydra lutris 166, **184**
Epomophorus wahlbergi 68, **51**
Equus africanus 234, **284**
Equus africanus somaliensis 234, **284**

Equus burchelli 241, **289, 290, 291**
Equus burchelli antiquorum 241, **289, 290**
Equus burchelli boehmi 241, **291**
Equus grevyi 237, **287**
Equus hemionus 235, **285**
Equus hemionus kulan 236, **285**
Equus kiang 237, **286**
Equus przewalskii 233, **282, 283**
Equus quagga 241
Equus zebra 237, **288**
Equus zebra hartmannae 237, **288**
Equus zebra zebra 237
Eremitalpa granti 58, **41**
Erethizon dorsatum 149, **160**
Erinaceus europaeus 60, **43**
Erinaceus roumanicus 61
Erythrocebus patas 110, **103**
Eschrithius robustus 336, **423.5**
Eubalaena glacialis 337, **423.6**
Eumetopias jubata 219, **261**
Euphractus villosus 121, **113**
Eutamias sibiricus 131, **127**

Felis chaus 200, **237**
Felis manul 200, **239**
Felis margarita 200, **238**
Felis silvestris 199, **235, 236**
Felis silvestris ornata 199, **236**
Felis silvestris silvestris 199, **235**
Fennecus zerda 195, **222**
Fossa fossa 187, **210**

Galago senegalensis 83, **72**
Galemys pyrenaicus 58
Galicitis vittata 162, **177**
Gazella granti 311, **380**
Gazella subgutturosa 310, **379**
Gazella thomsoni 310, **378**
Genetta genetta 182, **203**
Giraffa camelopardalis 286, **341, 342**
Giraffa camelopardalis reticulata 287, **341**
Giraffa camelopardalis rothschildi 288, **342**
Glaucomys volans 132, **129**
Glis glis 146, **153**
Globicephala melaena 339
Gorilla gorilla 114, **107**
Gorilla gorilla beringei 114
Gorilla gorilla gorilla 114
Gulo gulo 162, **178**

Halichoerus grypus 224, **268**
Helarctos malayanus 179, **199**
Hemiechinus auritus 61, **44**
Hemitragus jemlahicus 325, **399**
Hemitragus jemlahicus jemlahicus 325, **399**
Herpailurus yagouaroundi 199, **234**
Herpestes edwardsi 184, **207**
Herpestes ichneumon 184, **206**
Heterocephalus glaber 149, **159**
Heterohyrax syriacus 248, **295**
Hippopotamus amphibius 267, **316**
Hippotragus equinus 300, **357**
Hippotragus leucophaeus 301
Hippotragus niger 301, **358**
Hyaena brunnea 190, **214**
Hyaena hyaena 190, **213**
Hydrochoerus hydrochaeris 151, **161**
Hydrodamalis gigas 244
Hydropotes inermis 274, **323**
Hyelaphus porcinus 280, **332**
Hylobates lar 112, **104**
Hylochoerus meinertzhageni 263, **312**
Hypsignathus monstrosus 69
Hystrix leucura 148, **158**

Ictailurus planiceps 198, **229**
Indri indri 81
Inia geoffrensis 338, **415, 416**

Kobus ellipsiprymnus 308, **372**
Kobus ellipsiprymnus defassa 308, **372**
Kobus ellipsiprymnus ellipsiprymnus 308
Kobus kob 309, **373**
Kobus leche 309, **374**

Lagenorhynchus obliquidens 339, **419**
Lagostomus maximus 153, **163**
Lama glama 272
Lama guanicoe 271, **319**
Lama pacos 272
Lama vicugna 272, **320**
Lamprogale flavigula 162, **176**
Laptailurus serval 201, **240**
Lasiorhinus latifrons 41, **20**
Lemniscomys striatus 146, **152**
Lemur catta 80, **66**
Lemur mongoz 81, **67**
Leontopithecus rosalia 88, **77**
Leopardus tigrinus 198, **231**
Leopardus wiedi 199, **232**
Lepus americanus 259, **306**
Lepus europaeus 257, **304**
Lepus timidus 258, **305**
Litocranius walleri 310, **376, 377**
Lophiomys imhausi 139, **141**
Loris tardigradus 83, **70**
Loxodonta africana 251, **297, 298**
Loxodonta africana africana 252
Loxodonta africana cyclotis 253, **298 right**
Loxodonta africana oxyotis 253, **297, 298 left**
Lutra lutra 165, **183**
Lycaon pictus 197, **228**
Lynchailurus colocola 199, **233**
Lynx lynx 203, **241**
Lynx rufus 203, **242**

Macaca silenus 103
Macaca fascicularis 103
Macaca fuscata 103, **94**
Macaca mulatta 100, **93**
Macaca sylvanus 103
Macropus fuliginosus 48, **30**
Macrotis lagotis 41
Madoqua kirki 291, **346**
Manis pentadactyla 122, **116**
Marmosa murina 31, **8**
Marmota bobak 129, **124**
Marmota marmota 128, **123**
Marmota monax 130, **125**
Martes americana 159, **173**
Martes foina 160, **175**
Martes martes 160, **174**
Martes zibellina 159
Mazama americana 276, **325**
Megaderma cor 72, **56**
Megaleia rufa 47, **29**
Megaptera novaeangliae 336, **423.1**
Meles meles 164, **180**
Mellivora capensis 163, **179**
Melursus ursinus 177, **198**
Memphitis memphitis 165, **182**
Meriones meridianus 142, **146**
Mesocricetus auratus 139, **139**
Microtus arvalis 140, **143**
Miniopterus schreibersi 77, **61, 62**
Mirounga leonina 225, **270, 271**
Monachus schauinslandi 224, **267**
Moschus moschiferus 275
Mungos mungo 187, **208**
Muntiacus muntjak 275, **324**
Mus musculus 145, **149**
Muscardinus avellanarius 147, **155**
Mustela erminea 158, **169, 170**
Mustela nivalis 157, **168**
Myocastor coypus 154, **165**
Myotis myotis 77, **60**
Myrmecobius fasciatus 38, **16**
Myrmecophaga tridactyla 119, **109, 110**

Nasalis larvatus 97, **90, 91**
Nasua narica 171, **190**
Nasua nasua 171, **189**
Nemorhaedus goral 315, **385**
Neofelis nebulosa 206, **248**
Neomys anomalus 63
Neomys fodiens 63, **46**
Neotoma floridana 139, **138**
Nicrocebus murinus 81
Notoryctes typhlops 39, **17**
Nyctalus noctula 76, **59**
Nyctereutes procyonoides 195, **223**
Nycticebus coucang 83, **71**

Ochotona daurica 257, **303**
Ochotona princeps 257, **302**
Octodon degu 155, **166**
Odobaenus rosmarus 220, **262, 263**
Odocoileus hemionus 276, **326**
Oedipomidas oedipus 87, **75**
Okapia johnstoni 286, **340**
Ondatra zibethicus 141, **144**
Orcinus orca 341, **421**
Oreamnos americanus 315, **387**
Oreotragus oreotragus 291, **347**
Ornithorhynchus anatinus 28, **5, 6**
Orycteropus afer 123, **117**
Oryctolagus cuniculus 259, **307**
Oryx dammah 302, **361**
Oryx gazella 302, **362**
Otocyon megalotis 196, **226**
Ovibos moschatus 327, **401**
Ovis ammon 323, **398**
Ovis ammon ammon 323, **398**
Ovis ammon polii 323
Ovis canadensis 323, **397**
Ovis canadensis dalli 323, **397**
Ovis canadensis nivicola 323
Ovis musimon 319, **395, 396**

Pan troglodytes 115, **108**
Panthera leo 208, **252, 253, 254**

Panthera leo melanochaita 209
Panthera onca 208, **251**
Panthera pardus 206, **167, 249, 250**
Panthera tigris 211, **255, 256**
Panthera tigris altaica 213, **255**
Panthera tigris tigris 213, **256**
Papio anubis 104, **96**
Papio leucophaeus 108, **99**
Papio sphinx 107, **97, 98**
Paradoxurus hermaphroditus 182, **204**
Pardofelis marmorata 205, **246**
Pedetes capensis 137, **135**
Perodicticus potto 84, **73**
Peromyscus maniculatus 137, **136**
Petaurus breviceps 44, **24**
Petrogale penicillata 46, **27**
Petrogale xanthopus 47, **28**
Phacochoerus aethiopicus 263, **311**
Phalanger maculatus 42, **22**
Phascolarctos cinereus 40, **18, 19**
Phoca vitulina 221, **264, 265**
Phodopus roborovskii 139, **140**
Physeter macrocephalus 341, **412, 422, 423.8**
Pithecia monachus 93, **82**
Pithecia pithecia 91, **81**
Plecotus auritus 76, **58**
Pongo pygmaeus 112, **64, 106**
Potamochoerus porcus 263, **309, 310**
Potamochoerus porcus koiropotamus 263, **310**
Potamochoerus porcus porcus 263, **309**
Potamogale velox 54, **37**
Potos flavus 168, **186**
Praesorex goliath 65
Presbytis obscura 97, **89**
Presbytis pileata 97, **88**
Priodontes giganteus 121
Prionailurus bengalensis 198, **230**
Prionailurus euptilurus 198
Procavia capensis 248, **294**
Procavia johnstoni 248
Procyon cancrivorus 170, **188**
Procyon lotor 169, **187**
Profelis aurata 204, **244**
Profelis temmincki 205, **245**
Proteles cristatus 187, **211**
Pudu pudu 276, **327**
Puma concolor 205, **247**
Putorius eversmanii 159, **172**
Putorius putorius 159, **171**
Pusa sibirica 224, **266**
Pygathrix nemaeus 97, **92**

Rangifer tarandus 278, **329**
Rattus norvegicus 144
Rattus rattus 143, **148**
Redunca redunca 308, **371**
Rhinoceros unicornis 227, **272, 273**
Rhinolophus ferrumequinum 70, **53, 54**
Rhinolophus hipposideros 71, **55**
Rucervus duvauceli 283, **334**
Rupicapra rupicapra 315, **386**
Rusa unicolor 281, **333**

Saguinus imperator 88, **76**
Saiga tatarica 312, **383**
Saimiri sciureus 95, **86**
Sarcophilus harrisii 34, **13, 14**
Sauromys petrophilus 77, **63**
Sciurus carolinensis 126, **119**
Sciurus vulgaris 125, **118**
Selenarctos thibetanus 177, **197**
Sicista betulina 148, **157**
Sminthopsis crassicaudata 33, **11**
Solenodon cubanus 53, **36**
Solenodon paradoxus 54
Sorex alpinus 62, **35**
Sorex minutus 62, **45**
Spalax leucodon 142, **147**
Speothos venaticus 196, **225**
Suncus etruscus 61
Suncus murinus 64
Suricata suricata 187, **209**
Sus scrofa 263, **308**
Symphalangus syndactylus 112, **105**
Syncerus caffer 333, **411**
Syncerus caffer caffer 333, **411**
Syncerus caffer nanus 333

Tachyglossus aculeatus 25, **1, 2**
Tadarida brasiliensis 77
Tadarida teniotis 77
Talpa europaea 56, **39**
Tamandua tetradactyla 120, **111**
Tamias striatus 132, **128**
Tapirus bairdii 231
Tapirus indicus 233, **280, 281**
Tapirus pinchaque 232, **279**
Tapirus terrestris 231, **278**
Tarsipes spenserae 43
Tarsius syrichta 84, **74**
Taurotragus derbianus 298, **355**
Taurotragus oryx 298, **354**
Taxidea taxus 265, **181**
Tayassu tajacu 265, **314**
Tenrec ecaudatus 55
Tetracerus quadricornis 289
Thalarctos maritimus 181, **201**
Theropithecus gelada 103, **95**
Thylacinus cynocephalus 35, **15**
Tolypeutus tricinctus 122, **114**
Tragelaphus angasi 294, **350**
Tragelaphus imberbis 295, **351**
Tragelaphus scriptus 294, **348**
Tragelaphus spekei 294, **349**
Tragelaphus strepsiceros 295, **352, 353**
Tragulus javanicus 273, **322**
Tragulus meminna 273, **321**
Tremarctos ornatus 179, **200**
Trichechus manatus 244, **293**
Trichosurus vulpecula 42, **21**
Tupaia glis 79, **65**
Tursiops truncatus 341, **420**

Uncia uncia 215, **257**
Ursus americanus 177, **196**
Ursus arctos 173, **193–195**
Ursus arctos arctos 174, **193**
Ursus arctos horribilis 176, **194**
Ursus arctos middendorffi 176, **195**

Varecia variegata 81, **68**
Viverra civetta 181, **202**
Vulpes vulpes 192, **219**

Wallabia agilis 48, **31**
Wallabia parryi 50, **32**

Xerus erythropus 132, **130**
Xerus rutilus 132

Zaglossus bruijni 27, **3, 4**
Zalophus californianus 219, **260**